Nature's Choice
What Science Reveals About the Biological Origins of Sexual Orientation

Nature's Choice
What Science Reveals About the Biological Origins of Sexual Orientation

Cheryl L. Weill

Routledge
Taylor & Francis Group
New York   London

First published 2009
by Routledge
711 Third Avenue, New York, NY 10017, USA

Simultaneously published in the UK
by Routledge
2 Park Square, Milton Park, Abingdon, Oxon OX14 4RN

Routledge is an imprint of the Taylor & Francis Group, an informa business

Cover Design by Marylouise E. Doyle

Library of Congress Cataloging in Publication Data
   Nature's choice: what science reveals about the biological origins of
sexual orientation / Cheryl L. Weill
      p. cm.
   Includes bibliographical references and index.
   1. Sexual orientation—Physiological aspects. I. Weill, Cheryl L.
   QP81.6.N38 2007
   155.3—dc22

                                                             2007050849

ISBN10: 0-7890-3474-3 (hbk)
ISBN 10: 0-7890-3475-1 (pbk)
ISBN 10: 0-2038-8929-0 (ebk)

ISBN13: 978-0-7890-3474-8 (hbk)
ISBN 13: 978-0-7890-3475-5 (pbk)
ISBN 13: 978-0-2038-8929-9 (ebk)

This work is dedicated to:
The mothers, fathers, and family members of PFLAG,
who refuse to close their hearts and have found
the courage to open their minds.
&
The memory of Carl Sagan,
who believed that science should be accessible
to everyone.

CONTENTS

Foreword

The study of sexual orientation has experienced an explosion of exciting new discoveries in recent years, so many that they would be difficult to summarize in a single book. Nevertheless, Dr. Cheryl Weill has authoritatively covered the most important discoveries and integrated them into an overall understanding of why most people in all societies are heterosexual, while some are not. The book is both readable and technically accurate in depicting a fascinating array of old and new scientific knowledge.

Since Weill's book gives considerable attention to a theory that I and a former student formulated about 20 years ago (Ellis & Ames, 1987; Ellis, 1996), allow me to briefly sketch that theory here. It is called the *neurohormonal theory of sexual orientation* because its key premise is that human sexual orientation—whether homosexual, heterosexual, or somewhere in between—is largely determined by hormones operating on the brain. More specifically, **sex** hormones (especially testosterone), operating on primitive parts of the brain in and around the limbic system guide the development of sexual orientation. This means that the higher brain centers used in rational thought and language comprehension have no direct bearing on sexual orientation.

Theoretically, the most crucial time for hormonal influences on sexual orientation is during the first six months after conception. In other words, sexual orientation is not learned, it is determined prenatally, although it will not be expressed in the form of sexual behavior and preferences until sexual maturity. If high levels of testosterone or one of its metabolites are prevalent in and around the limbic region of the brain prenatally, an individual's adult preference will be for female sex partners. If low, the preference will be for male sex partners.

What determines the availability of testosterone prenatally? Genes are very important, particularly ones located on the Y-chromosome

Nature's Choice

(which only males have). These genes virtually ensure that would-be ovaries will form into testes instead. Once this happens, the testes become efficient little factories for making testosterone, but many other factors can complicate the process.

As Weill describes very well, considerable evidence suggests that the intrauterine environment can sometimes partially override the genetic instructions in several ways. For example, stress hormones secreted by the mother can temporarily shut down testosterone production by the fetus's testes. If this occurs during critical periods of brain development, sexual orientation can be altered from its normal course. Similarly, various drugs taken during pregnancy can either block or augment the fetus's production of testosterone or divert it from its usual metabolic pathway. Also, still poorly understood immunological factors appear to alter how sex hormones guide sexual orientation. Finally, the genetic instructions themselves sometimes mutate in ways that redirect sexual orientation during fetal development.

Overall, the neurohormonal theory maintains that a person's sexual orientation is the result of a complex and delicately balanced process whereby sex hormones operate on the brain during fetal development to ensure that most humans have offspring so as to perpetuate the species. A simple way to think about the theory is that it asserts that some of us, through no fault of our own, have brains that were "sexed" in ways that are contrary to how our genitals were "sexed."

The neurohormonal theory stands in opposition to the idea that sexual orientation is learned through socialization. Although the theory recognizes that learning is a powerful force in affecting the expression of sexuality, it maintains that the major forces behind our basic preferences for either male or female sex partners is determined long before we are culturally socialized.

Weill's book is useful in providing a well-documented summary of research findings on a wide array of traits that statistically distinguish heterosexuals and homosexuals. To give just a few examples, Weill discusses several studies that have found male homosexuals having larger penises than do their heterosexual counterparts. And, with few exceptions, certain finger-length ratios have been repeatedly found to distinguish males from females and heterosexuals from homosexuals within both sexes. Research that I find particularly interesting is that which statistically links left-handedness with homosexuality (partly because I am a left-hander, albeit heterosexual). Weill reveals how these and several other fascinating average differences could be related to testosterone's complex effects both inside and outside of the brain.

In conclusion, owing partly to her background in biochemistry, Weill has assembled an impressive document on what science now knows about the causes of variations in sexual orientation. And, the author's training in social work has ensured that, throughout the book, readers never lose sight of the human dimension in sexual orientation. This book should be read by everyone who is truly interested in this intriguing aspect of human sexuality.

Preface

The biological origins of sexual orientation became an intense focus of the media in the early 1990s with the publication of two seminal papers in the scientific journal *Science*. The first, by Simon Le Vay, titled "A Difference in Hypothalamic Structure between Heterosexual and Homosexual Men," appeared in 1991. The second, from a group led by Dean Hamer, titled "A Linkage between DNA Markers on the X Chromosome and Male Sexual Orientation," appeared in 1993 (Hamer, Hu, Magnuson, Hu, & Pattatucci, 1993b). At the time, I was an associate professor in the departments of Neurology and Anatomy at Louisiana State University Medical School in New Orleans. Although I have a PhD in organic chemistry, I had spent the previous 20 years doing basic research in the area of cellular and molecular neurobiology. Thus, I found both the anatomical and genetic data regarding sexual orientation intriguing. In the summer of 1993 I was approached by Dr. Norma Kearby, a board member of the local chapter of Parents, Families and Friends of Lesbians and Gays (PFLAG), with the idea of giving a lecture on the biology of human sexuality at the upcoming national PFLAG conference. She asked that I pay particular attention to Le Vay and Hamer's recent data on the neuroanatomy and genetics of sexual orientation. As is often the case, at first thought it seemed like a simple request, so I agreed. I soon realized that I had a great deal to learn in order to successfully explain the latest anatomical and genetic findings to a general audience. Fortunately, Lee Ellis and M. Ashley Ames had published an excellent comprehensive review in 1987 titled "Neurohormonal Functioning and Sexual Orientation: A Theory of Homosexuality–Heterosexuality" (Ellis & Ames, 1987). My approach was to search the scientific literature from 1987 on for all papers that even mentioned biology and sexuality or sexual orientation. After considerable study I was able to put together what I think is an accurate presentation of the data that support the biological origin of human sexuality, including sexual orientation. The lecture was well received at the national PFLAG convention, and as a result I have since given the lecture extensively to other PFLAG chapters as well as other service organizations and graduate and medical schools. As more data became available, it was clear that a single lecture was not enough. I could not reach enough people with all of the relevant information with a lecture, hence the need for something more substantial in written form.

In assembling the material for the lecture and the book, I was struck by the fact that much of this material has been in the scientific literature for 20 years or more, but has yet to be directly incorporated into human sexuality courses as taught at the secondary and college levels. I hope this material will find its way into these courses. It is my firm belief that knowledge, no matter how sophisticated the methods used to obtain it, should be accessible to everyone.

The material presented is much more than is possible to give in a single hour lecture. I believe the added material provides additional support for the biological origins of sexual orientation, and as such broadens one's understanding of human sexuality. Glossary terms appear in bold in text at first mention, and I have also included an appendix that contains additional information that expands on the material in the text.

This material is written for people who have little or no background in science. In the interests of clarity, the information is presented in a simple manner that I hope still conveys the information accurately. Distilling the most important information from the primary scientific literature is a challenging process. I am indebted to all the scientists that have contributed to our current understanding of sexual orientation. Any errors or misrepresentations of the material are solely my responsibility. People with a scientific background will undoubtedly find the presentation wanting, and I refer them to the cited literature for a more in-depth presentation of the material.

Acknowledgments

This book is the end product of a lecture that I presented at the 1993 National PFLAG Convention in New Orleans. I did so at the request of Dr. Norma Kearby. I thank her for having put me on the path that has led to this book. Over the years a number of PFLAG folks have doggedly kept me on that path. Most notably, I am indebted to PFLAG Houston, Texas, for their collective support and encouragement. I am also very grateful to Dr. Jerry Bigner for seeing value in a book that tries to make complex scientific concepts accessible to general audiences. I thank Magdalen Treuil, my steadfast technician for 20 years, for proofreading the initial version of the manuscript. Last, I thank my partner Judy Calhoun, for her patience and help proofreading all of the final chapters.

Chapter 1

Introduction

Nature or nurture? Is human behavior dictated by biology or determined by our environment? Nowhere in this country today is this question more hotly debated than with respect to **sexual orientation.** Is our sexual orientation inherent, or is it influenced by our social environment and therefore something that is learned? Some people claim that sexual orientation is a choice and learned through environmental influences. This belief, infused with the belief that the only morally right choice is heterosexuality, has and continues to spawn hate and violence toward **homosexuals.** This position has political dimensions as well, as reflected in the U.S. government's handling of homosexuals in the military through a policy of "don't ask, don't tell." Furthermore, gay people work, pay taxes, and can serve in the military and die defending this country, but in most states cannot marry the person they love.

Alternatively, it is also argued that sexual orientation is inherent, that is biologically determined. Thus, we don't have a choice about what we feel, but we do have a choice about whether and how we act on those feelings. This is true regardless of orientation. Which position is correct? Are there scientific data to support one or the other position? The answer to the last question is yes. Considerable data in the scientific literature support the conclusion that human sexual orientation is determined during gestation, is highly dependent on available **testosterone,** is fixed at birth and cannot be changed after birth, and is independent of any identifiable family and cultural environmental influences. In contrast, arguments and theories of how social environment contributes to human sexual behaviors have persisted in the absence of any supporting credible scientific studies. Why are we as a society seemingly so unaware of these facts?

Nature's Choice

There are at least two primary explanations. First, human sexual development is not taught in a way that includes the origins of sexual orientation. This is generally true at all levels of education, from high schools, through colleges and universities, graduate schools, and professional schools. Second, although some books present some of the relevant data, these presentations do not focus on the origins of sexual orientation. Furthermore, those books that do present some of the relevant data on homosexuality are too broadly or narrowly focused and are not written for general lay audiences. What is needed is a focused presentation written for lay audiences that describes the scientific data pertaining to the determination of sexual orientation, including heterosexuality, bisexuality, and homosexuality.

Lee Ellis and M. Ashley Ames published an excellent comprehensive review in 1987 titled "Neurohormonal functioning and sexual orientation: A theory of homosexuality—heterosexuality" (Ellis & Ames, 1987). *Nature's Choice: What Science Reveals About the Biological Origins of Sexual Orientation* describes human sexual development and presents a logical and sequential survey of the scientific data published from 1987 through early 2007 identifying the biological molecules and processes that appear to determine both **heterosexual** and homosexual orientation in humans.

The book is divided into two sections. Section I presents some basic definitions and facts about sexual orientation followed by a succinct description of human sexual development. Next, the effects of **hormones** on sexual development are described. This in turn provides the background for presentation of the prevailing theory of the biological determination of sexual orientation developed by Ellis and Ames: the gestational neurohormonal theory. Section II defines science, and introduces experimental design scientific data terminology and interpretation as a prelude to the presentation of experimental results. Also included is an overview of how scientific research is funded in the United States. Next, descriptions of the results of studies of all of the areas examined thus far with respect to sexual orientation are presented. These include brain anatomy, genetics, **sex**-typical behavior in children, auditory, startle reflex, and **olfactory** responses, anthropometrics or body measurements and physiology, cerebral lateralization, handedness and **cognitive** abilities, maternal stress and substance use, and gay men with older brothers. A conclusion is presented based on all of the results. Last, a speculative model of the role of testosterone in determining human sexuality is offered. Also included are many figures and tables, a glossary, and an appendix. A Recommended Reading section is included, and references to original studies are provided in the Bibliography.

I have tried to present the material so that it is readily understandable to as broad an audience as possible. I hope I have succeeded. I have also included a discussion of the emotional basis of sexual attraction and behavior that I hope is illuminating. It is my foremost hope that you will come away with a clearer understanding of what science can tell us about sexual orientation, as well as what science cannot tell us. With this knowledge, I leave it up to you to decide how you personally feel about homosexuality, to what degree it is influenced and/or determined by biology, and thus to what extent it is a matter of personal choice.

Chapter 2

Definitions

Before introducing the biology of sexual orientation, let us look at some pertinent facts and definitions.

HOMOSEXUALITY IS NOT A MENTAL DISORDER OR DISEASE

One of the most important things to know about homosexuality is that it is not a mental disorder or disease. In 1973, the American Psychiatric Association removed homosexuality from its official *Diagnostic and Statistical Manual of Mental Disorders,* signifying the end to its official classification as a mental disorder or disease. It might be helpful here to specifically define what we mean by disease. A disease is defined in *Webster's Ninth New Collegiate Dictionary* as "a condition of the living animal body, or of one of its parts that impairs the performance of a vital function." A vital function is one required to sustain the life of the individual. Homosexuality does not impair any human vital function, and thus, in this sense also, it is not a disease.

GENDER CHARACTERIZATION

How do we characterize a person's gender, or whether they are male or female? Science distinguishes between male and female sexuality in six ways, three physical and three behavioral. First, a distinct difference exists in the **chromosomes** possessed by males and females. Most of us are born with either of two **sex chromosome** pairs:

Nature's Choice

XX if you are female and XY if you are male (see Appendix for a discussion of fertilization and sex chromosome determination). Second, our **gonads,** or primary sex glands, differentiate into either testes or ovaries. Third, our sex is determined by the appearance of our external **genitalia.** There is every expectation that if your genitalia look male, that is, a **penis** and **scrotum** are present, your sex chromosomes are XY and you possess **testes.** Similarly, if your genitalia look female, a **vagina** and **clitoris** are likely within labia, your sex chromosomes are XX, and you possess **ovaries** (we will examine situations in which these conditions are not met in Chapter 4).

There are three behavioral traits that we generally consider when we speak of ourselves as female or male. First, we make a distinction between how we see and feel about ourselves internally (that is, in our minds). For most of us, our mental image of and feelings about ourselves match our physical gender. However, for a small number of individuals, their internal image and feelings about themselves are opposite to their physical gender. Such individuals often seek to have their physical gender changed, either male to female or female to male. They are referred to as transgendered or transsexual persons. Scientists do not have a great deal of information on the neurological origins of this condition, although some studies have been published (see Zhou, Hofman, Gooren, & Swaab, 1995). This will not be discussed further, as it is not clear how the biology of sexual orientation relates to the transgender state. Second, we make a distinction based on sexual orientation. The expectation is that you will orient toward the opposite sex (that is, you will be heterosexual). A small percentage of the population orient toward members of the same sex and are homosexual. Some individuals are attracted to both men and women and are thus bisexual. Third, we distinguish between what we call **male-typical** and **female-typical** behavior, referred to as sex-typical behavior, or gender-role behavior. Once again, if you appear male, there is every expectation that you will display a range of male-typical behaviors. Similarly, if you appear female you will be expected to display a range of female-typical behaviors. Sex-typical behaviors are easily appreciated in young children. For example, rough-and-tumble play and playing competitive games are considered male-typical behaviors, while maternal imitative behaviors and playing dress-up and house are considered female-typical behaviors. There are also female- or male-typical ways of talking, gesturing, sitting, standing, and walking. Gate, or the way we walk, is an interesting behavior, which can appear distinctly feminine or masculine. No one teaches us how to walk; we just do it. It is a personal attribute that develops outside of our cognitive awareness. However, we can teach ourselves to walk differently if we want to. That's because, as we will see in Chapter 3, sex-typical behaviors are associated with diverse areas of the brain, extensively involving the **cortex,** or our thinking brain; we use our cortex in order to walk. Therefore, we can use our cortex in conjunction with other areas of our brain to learn a new way of walking.

Many of us are not aware and still others tend to ignore or deny that real variations in these gender traits do exist. We know that a relatively small number of individuals are born with more than two sex chromosomes. We also know that a small number of individuals are born with external genitalia that are ambiguous, that is, they do not appear distinctly male or female, or do not correlate with their sex chromosomes. Most of us are generally tolerant of a wide range of variability in sex-typical behaviors. However, considerable ignorance of and a substantial resistance to understanding sexual orientation variability exists, be it bisexuality or homosexuality. In general, human gender and sexuality are determined at birth (sex chromosomes and external genitalia appearance) and subsequently discovered by all of us, usually in adolescence or adulthood (sex-typical behaviors and sexual orientation).

SEXUAL ORIENTATION

What do we mean by sexual orientation? Standard dictionary definitions of sexual orientation include: (1) the direction of sexual feelings or behavior toward the same sex, opposite sex, or some combination of the two; and (2) consistent preference for sexual relations with one's own sex (homosexuality), opposite sex (heterosexuality), or varying degrees of ambivalence about the partner's sex (bisexuality). These are accurate definitions of what we do physically, but they provide no insight into the psychological and emotional motivations for human sexual interactions. In short, they define what we do, but not why we do it, whether we are heterosexual, homosexual, or bisexual. So, what is the motivation behind the expression of our sexual orientation? What are the feelings that precede the physical expression of our sexuality?

Most of us can agree that in order to express our sexuality, we need to experience an emotional response to a person, some kind of affection, fondness, or love. This is true regardless of sexual orientation. So, if an emotional response or feelings for a person precedes the expression of our sexuality, this suggests an interesting retrospective thought experiment that everyone can do to determine his or her own sexual orientation. The experiment: Think back to about the time when you were entering **puberty,** usually early teens. It could be a little earlier or later, but it is usually about this time when most of us have this experience for the first time. Identify for yourself the first person that you just could not take your eyes off of, you felt wonderful just being around him or her, they just made you feel calm, complete, and glad to be alive. This person captivated you. It was the beginnings of falling in love for the first time. Who was that person? More important, what was their gender? I submit to you that the person's gender defines your sexual orientation, or at least part of it if you are bisexual. In general, about 96 percent of boys experience this in response to a girl, while about 98 percent of girls experience this in response to a boy. However, for 2 to 4 percent of boys it is in response to another boy, and for about 1 to 2 percent of girls it is in response to another girl. (A recent examination of sexual orientation among 8,000 college students from the United States and Canada found 97 percent of participants labeled themselves as heterosexual, with the proportion of homosexuals and bisexuals combined being 3 percent of the male and 2 percent of the female participants [Ellis, Robb, & Burke, 2005].) This first experience, or what we could call the prelude to "falling in love," is usually so profound that most people can remember not only the sex of the individual but also his or her name and the life circumstances surrounding this first awareness of being so charmed and bewitched by another human being.

Now there are a couple of interesting facets to this experience. First, ask yourself, did this awareness contain or was it associated with any blatantly sexual feelings? For most people the answer is no, and most admit that at that time in their lives they either knew nothing about sex, or knew about it but had no awareness of it in the context of the feelings associated with recognizing this special person. I suggest that sexual feelings are secondary to and occur after the recognition of this person. We only learn to associate potential sexual feelings with this recognition experience as we mature. So, we might say that the natural temporal sequence is feelings of infatuation first followed by sexual attraction or feelings.

Second, did you pick this person? That is, did you know you wanted to have such feelings for someone and then look around and decide that this was the right person? Once again, for most people the answer is no. We have no prior knowledge that we are going to have this experience and there was no choice, or for that matter no conscious thinking involved in the experience. Usually the awareness of this person comes to us out of the blue, totally unanticipated and unexpected. One could say that life brings this experience to us. In fact, this experience is one of life's more wondrous mysteries. Neither psychologists nor psychiatrists understand it or can explain it. However, most of us do agree that you can't select this person using logic and reason; you can't *make* it happen. You just have to wait until your heart (actually, it is the brain) recognizes that special person.

An alternative way of recognizing or defining one's own sexual orientation is the "directional marker," attributed to Richard Pillard (Burr, 1996). "Sexual orientation is, when you pass people on the sidewalk at lunch hour, which sex do your eyes flick on involuntarily: men or women?" (Burr, 1996) The important aspect of this observation is that it occurs involuntarily. You are not thinking about any aspect of your own sexuality; you are just lost in your own thoughts when this happens.

An alternative to the term sexual orientation suggested by Dr. John Money is "amative orientation" (Money, 2002). Amative signifies not only sexual arousal and practices, but also the corresponding ideation, imagery, dreams, reveries, fantasies, and cognitive rehearsals.

Here is another thought experiment that can tell us something about the degree to which we exercise choice in our sexual interactions. Once again, think back to your first sexual experience with someone whom you felt you truly loved. Now go to the time during which you and this person were approaching the decision to have sex. Usually, we are at that point in the relationship where no words are left to express how we feel. Most of the communication between us and our beloved is through eye contact and other means that we cannot define. Now, the question is, what are you thinking about and to what degree do your thoughts influence your actions? Most people can recall that they were thinking about many things, particularly why they shouldn't do what they were about to do, and foremost their parent's prohibitions and cautions. In some cases they even remember being aware that their actions could lead to a pregnancy and that that was not good or desirable. The next question is, did you act on any of your thoughts? Usually the answer is not at all or very little. Why is that? I submit that the communication that is taking place at that time by whatever means is so compelling that very few considerations, that is, reasoned thoughts, are allowed to interrupt it. This is evident from the number of such encounters that occur across strongly prohibited boundaries imposed by family, culture, and society, to say nothing of the various and sundry places where it occurs. This is a strange situation; we have choices, but we just do not exercise many of them during that first loving sexual experience, and this is true whether the person across from you is the same sex or the opposite sex. Once again, we are not exercising choice, but are compelled to act, and the sex of the person with whom we are so compelled to communicate sexually is of little consequence. Sex per se is not the primary consideration, but rather continued communication and the completion of that communication are the goals. It is just that sex is the activity that nature has given us to achieve these goals.

So, where does this leave us concerning the definition of human sexual orientation? It seems that the feelings that dictate sexual orientation are not sexual, but emotional, and they are not a matter of choice. Furthermore, it seems the specific person that will elicit these feelings is unique to each individual and cannot be predicted, even by the individual. Therefore, no one can tell us what our sexual orientation will be. Each of us must discover it for ourselves through our unique emotional experiences. The universality of this experience can be inferred from its persistent appearance in literature down through the ages. The experience doesn't seem to have changed very much for human beings for thousands of years, and it remains the most mysterious and compelling aspects of our personal and interpersonal lives. We can conclude, therefore, that one's sexual orientation, whether heterosexual, homosexual, or bisexual is not a choice and is discovered by each of us through our personal emotional experiences. We will explore the degree to which science supports or refutes these observations in the remaining chapters.

TWO MYTHS

Two myths are often brought up in the context of homosexuality. The first is that homosexuality is pedophilia. Pedophilia is defined as sexual perversion in which children are the preferred sexual object. Also, pedophiles can have the same range of sexual orientations as the general population, from heterosexual to homosexual. No evidence suggests that any relationship exists between homosexuality and pedophilia (see also Appendix).

The second myth concerns the idea that sexual abuse during childhood can alter or influence one's sexual orientation. As we will see, the available science suggests that one's sexual orientation is determined prior to birth and is not alterable. Therefore, sexual abuse in childhood, whether by an adult male or female, should not alter or influence one's sexual orientation. It is clear, however, that childhood sexual abuse does impact a person's ability to express his or her sexuality as an adult (see also Appendix).

SUMMARY

In summary, sexual orientation is not a mental disorder. We see that gender, that is, whether one is male or female, can be characterized by three physical characteristics and three behavior characteristics. Physically, females have XX chromosomes and males have XY chromosomes. Males have testes, and females have ovaries. Last, males and females have distinctly different external genitalia; males have a penis and a scrotum, while females have a vagina and a clitoris within labia. Behaviorally, we have a gender identity, whether we feel ourselves to be male or female; a sexual orientation, whether we are attracted emotionally and physically to same-sex or opposite-sex individuals; and sex-typical behavior, whether we display same-gender or opposite-gender behaviors.

Sexual orientation—heterosexual, bisexual, and homosexual— have two components, an emotional component and a physical component. For most of us, it seems that the emotional component either precedes the physical response or is integrated with the physical response in time, we do not pick the person we are attracted to, nor are we focused on sex specifically when we identify someone that we are attracted to.

Furthermore, homosexuality is not pedophilia, and no evidence shows a relationship between homosexuality and pedophilia. Last, childhood sexual abuse does not lead to a homosexual orientation, although it usually does interfere with the expression of one's inherent sexuality regardless of sexual orientation.

Chapter 3

Human Sexual Development

Sexual development occurs in two phases. The first organizational phase occurs during gestation and involves the development of the internal and external genitalia and most or all of the brain circuits that provide for adult sexual behavior. The second, or activational, phase of sexual development takes place at puberty and throughout adulthood, when under the influence of sex hormones, both male and female anatomy mature to their full adult form and function. The brain circuits that support adult sexual behavior also become fully functional under the influence of the pubertal surge of sex hormones. How sex hormones influence these processes will be examined in Chapter 4.

First, let's examine how the morphological development of our bodies and the neurological development of our brains proceed during gestation.

MORPHOLOGICAL DEVELOPMENT

Morphological development involves two areas, the internal and external genitalia (Figure 3.1).

Internal and External Genitalia

Male

In males, the internal genitalia are the epididymus, vas deferens, prostate gland, and the seminal vesicle. The external male genitalia are the penis and scrotum.

Nature's Choice

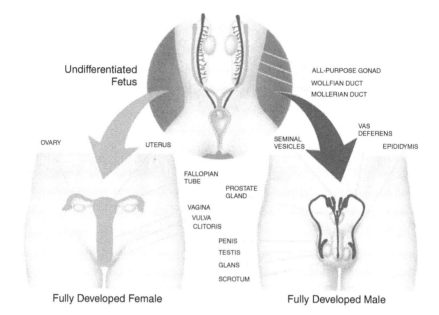

FIGURE 3.1. Normal gender progression: In normal gender development, an all-purpose gonad develops either as an ovary or as a testis depending on the combination of chromosomes present. With an X chromosome from both the male and female parents, it develops as an ovary; with an X from the female parent and a Y from the male parent, it develops as a testis. In males, the testes produce hormones, or androgens, that convert certain embryonic structures into the appropriate male parts. Without the influence of these androgens, the same structures normally develop into the female counterparts. *Source:* Diamond (1992).

Female

In females, the internal genitalia are the oviducts, **uterus,** and cervix, while the external genitalia are the vagina, **labia minora, labia majora,** and clitoris.

Gender Progression

In this section I will explain how these structures come about during gestation. The first and I think most intriguing fact is that prior to the seventh week of gestation, no observable physical difference exists between an undifferentiated male and female fetus; they look the same externally and possess the same structures internally (Figure 3.1). Both possess bi-potential gonads that can give rise to either ovaries or testes. Both possess **Müllerian** and **Wolffian ducts** that can give rise to either the internal female or male genitalia respectively. In addition, each possesses structures that can give rise to the female and male external genitalia.

So, what determines which structures will develop? Human cells contain 23 pairs of chromosomes on which all of the genes necessary for development and life are located. For 22 of these sets of chromosomes, the pairs are identical. The 23rd pair, the sex chromosomes, can be either identical (XX), indicating a female, or nonidentical (XY), indicating a male. If the sex chromosomes are XX, then around the 13th week of gestation the gonads develop into ovaries, the Wolffian ducts (precursor of the male internal genitalia) atrophy and disappear, and the Müllerian ducts develop into the female internal genitalia, including a uterus, fallopian tubes, and inner vagina. In addition, the external structures develop into the female external genitalia including clitoris, labia minora, and labia majora.

A much different outcome results if the sex chromosomes are XY. Genes on the Y chromosome code for two factors that switch the development of these same structures to the corresponding male forms. The first of these is **testis-determining factor (TDF)**, which directs the gonads to become testes on about the eighth week of gestation. During the ninth week of gestation the testes become active and start to produce the steroid hormone testosterone, which directs the development of the Wolffian ducts into the male internal genitalia, which includes the seminal vesicles, vas deferens, and epididymus. Testosterone is also converted to **dihydrotestosterone,** or DHT, by the **enzyme** 5-alpha-reductase. Under the influence of locally produced DHT, external precursor structures are transformed into the external male genitalia including the penis shaft, glands penis, and scrotum. Another **gene** on the Y chromosome codes for another factor called **Müllerian inhibition hormone** (MIH), which prevents the further development of these ducts into the internal female structures. The Müllerian ducts atrophy and disappear.

The overall picture that emerges is that in the absence of any hormonal influence, the fetus will develop female external and internal genitalia. In order for the fetus to develop the corresponding male genitalia, instructive factors coded for by genes on the Y chromosome and testosterone must be available. Bottom line: do nothing, and in the absence of testes determining factor, testosterone, and Müllerian duct inhibiting hormone, the fetus will become female, regardless of the sex chromosomes. To get a male fetus requires additional factors and hormones.

There are two additional facts that we will explore in more detail later: (1) a male fetus (possessing XY chromosomes) that is not exposed to testosterone or is unable to respond to testosterone will become female, and (2) a female fetus (possessing XX chromosomes) that is exposed to male levels of testosterone will become male. This is indicated diagrammatically in Figure 3.2.

Last, we must remember that during development and throughout life, testes produce primarily testosterone, but also low levels of **estrogen.** Conversely, ovaries produce primarily estrogen and progestins, but also low levels of testosterone.

Hormonal sex determination

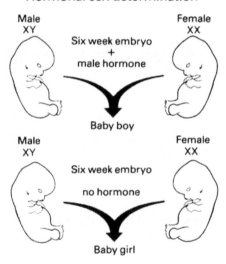

FIGURE 3.2. Hormonal sex determination: Androgen hormones influence the development of male (XY chromosomes) and female (XX chromosomes) fetuses. If adequate androgen hormones are present, both a male and female fetus will develop along the male path and become a boy. If no androgen hormones are present, both a male and a female fetus will develop along the female path and become a girl. *Source:* Moir & Jessel (1991).

Genitalia Development in Utero

One can see in Figure 3.3 how the identical undifferentiated external genital structure diverges during development, assuming either the male or female form. One characteristic that can be seen in this figure is the size of the anogenital space. This is the distance between the anus and the **urethra,** which increases during development and is greater in adult males than in adult females. The anogenital distance is often used as an index of the degree to which a female has been masculinized during development, as we will see later (see Chapter 4).

NEUROLOGICAL DEVELOPMENT

While the body of the fetus is developing its sexual characteristics, the nervous system is also developing. Specific areas of the nervous system develop in stages, are influenced by steroid hormones, and sequentially determine sexual orientation and sex-typical behaviors.

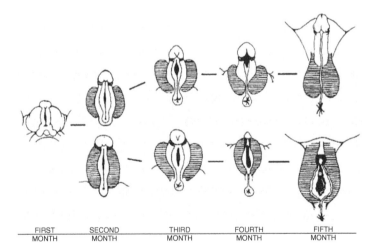

| FIRST MONTH | SECOND MONTH | THIRD MONTH | FOURTH MONTH | FIFTH MONTH |

FIGURE 3.3. Genitalia development in utero: As a fetus grows, there are two possible ways the external sex organs can develop. The top figures are male structures and the bottom figures are female structures. *Source:* Silber (1981). From *The Male from Infancy to Old Age* by S.J. Silber, 1981. Reprinted with permission of Thomson Learning: www.thomsonrights.com. Fax 800-730-2215.

To understand how the brain centers involved in sexuality relate to and function within the entire brain, it is instructive to view brain development and function from an evolutionary perspective.

Paul MacLean, the former director of the National Institute of Mental Health, has characterized the human brain as a "triune brain," that is, composed of three evolutionary distinct areas (MacLean, 1990). The first brain system that developed, and therefore the most primitive, is the reptilian brain, or "R-complex." It includes the brain stem, cerebellum, and other structures. The behavior of lower vertebrates such as lizards and snakes is dominated by the brain stem and cerebellum. It is characterized by routines or programs and is rigid, obsessive, compulsive, ritualistic, and paranoid. Their behavior is repetitive and they do not learn from past experience. The R-complex is concerned with survival and body maintenance and controls muscles, balance, and autonomic functions such as heart rate and breathing.

The second area of the brain that evolved is the limbic system, or the middle part of the brain, and is often referred to as the "old mammalian brain" since it first appeared in the earliest mammals. It is concerned with emotions and instincts, feeding, fighting, fleeing, and sexual behavior. The limbic system as a whole is the primary seat of emotion and affective memories, or emotionally charged memories. The brain structures that make up the limbic system are the **hypothalamus,** hippocampus, and amygdala. The limbic system helps you know whether you feel positive or negative about something and what gets your attention. As you might expect, it is extensively interconnected with the neocortex, or cortex, our reasoning brain.

MacLean (1990) has suggested that the limbic system provides the biological basis for the tendency of thinking to be subordinate to feelings, to rationalize desires. This lowly mammalian brain of the limbic system tends to be the seat of our value judgments. It decides whether our higher brain (or cortex) has a good idea or not, and whether that idea feels true or right to us.

The last area of the brain that evolved is the neocortex, also referred to as the cerebral cortex, or just cortex. It is our superior or rational brain and the seat of higher cognitive functions that are unique to humans. The cortex comprises almost the whole of the hemispheres and some subcortical structures and corresponds to the brain of primate mammals, including humans. MacLean refers to the cortex as "the mother of invention and the father of abstract thought." The human cortex accounts for about 85 percent of the total brain mass. Most animals have a rudimentary cortex, with mammals having the most complex cortex. Primates have the most evolved cortex, which is essential for function. Disconnect the cortex of a mouse and the mouse appears to be able to do mouse things perfectly well; disconnect the cortex of a human being and they don't move and can't communicate.

The cortex has left and right hemispheres; the right is more spatial, abstract, musical, and artistic, while the left is more linear, rational, and verbal. It is the seat of language, formal operational thinking, and makes planning possible.

Brain Areas Related to Sexual Orientation

Some scientists think sexual orientation is related to the establishment of permanent differences in the limbic areas of the brain, particularly the hypothalamus. As we said, the limbic system helps you know whether you feel positive or negative about something and what gets your attention. The limbic system is extensively interconnected with the cortex, but it is not clear if the cortex is involved in the determination of sexual orientation (see Figure 3.4).

Brain Areas Related to Sex-Typical Behaviors

In contrast, sex-typical behavior patterns are associated with diverse areas of the brain, extensively involving the cortex (Figure 3.4).

SUMMARY

The two areas of sexual development are morphological and neurological. Morphological development includes the internal and external genitalia. Males and females have unique internal and external genitalia. Prior to the seventh week of gestation, the fetus is undifferentiated and has the capacity to become either male or female regardless of which chromosome pair it possesses. Male development proceeds when testis-determining factor is produced. This factor instructs the undifferentiated gonads to become testes. The testes pro-

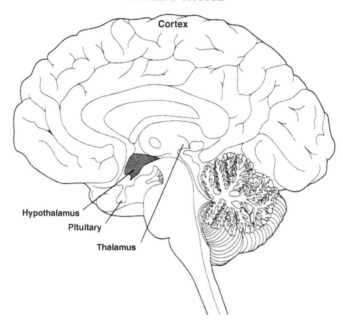

FIGURE 3.4. Cross-sectional view of the human brain: Medial view of the human brain showing the relationship of the hypothalamus to the pituitary and **thalamus.** *Source:* Kandel, Schwartz, & Jessel (1995). Kandel, E.R., Schwartz, J.H., & Jessell, T.M. (Eds.), *Essentials of Neural Science and Behavior.* Copyright 1995. Appleton & Lange. Reproduced with permission of The McGraw-Hill Companies.

duce testosterone, which drives most of the rest of male development. If testes determining factor is not produced, the gonads will become ovaries on about the thirteenth week of gestation, and the fetus will develop as female. If a male fetus is not exposed to testosterone, it will develop as a female. Conversely, if a female fetus is exposed to male levels of testosterone, it will develop as a male. Last, ovaries produce estrogen and also low levels of testosterone. Similarly, testes produce primarily testosterone, but they also produce estrogen.

While the body is developing, the nervous system is also developing in stages and is influenced by steroid hormones, especially testosterone. It is thought that sexual orientation is related to establishment of permanent differences in the limbic areas of the brain, particularly the hypothalamus. Although the limbic system is interconnected with the cortex, it is not clear if the cortex is involved in the determination of sexual orientation. On the other hand, sex-typical behavior patterns are thought to be associated with diverse areas of the brain, extensively involving the cortex.

Chapter 4

Hormones and Sexual Development

Before we look at the effects of hormones on sexual development, it would be helpful to understand how steroid hormones work and how the brain uses them. The steroid hormones of interest are the gonadal hormones produced by the gonads (that is, the ovaries and testes) including estrogen, **progesterone,** and testosterone. An additional source of related **steroids** is the adrenal glands. Two features of gonadal hormone action in the brain support the development of sexual behavior: how brain cells, both **neurons** and non-neuronal cells, use or metabolize them, and how we think they work in these cells to provide for behavior. It will be useful to understand these ideas before looking at sexuality in normal animals, animals under experimental conditions, and in related human conditions. Because experimentation on human beings is unethical, scientists use studies of normal animals and the results of animal experiments to give them insights into how hormones influence adult sexual behavior. We will see that animal studies have helped us understand the timing of hormonal action and the effects of the presence and absence of testosterone on both male and female sexual development.

GONADAL HORMONES AND HOW THEY WORK

How the Brain Uses Gonadal Hormones

To understand how the brain is sexually organized during the critical period of development we need to know what happens to steroid hormones when they reach the brain. The **androgen,** testosterone (T) is the major steroid hormone modified by brain cell enzymes. There are two metabolic pathways in neurons and non-neuronal cells for T. One pathway requires the enzyme 5-alpha-reductase, which converts T to dihydrotestosterone (DHT). The conversion of T to DHT in the brain takes place primarily in cells of the **pituitary** and brain stem. Outside of the brain, DHT is required for the development of the male external genitalia. The other pathway for T metabolism requires the enzyme **aromatase,** which converts T to 17 beta-estradiol (E), an estrogen. The conversion of T to E in the brain occurs mostly in neurons of the hypothalamus and limbic system.

Organizational and Activational Effects of Steroid Hormones

In 1959 a hypothesis was presented to explain the effect of steroid hormones on the development of sexual behavior, especially in mammals (Phoenix, Goy, Gerall, & Young, 1959). The hypothesis states that steroid hormones affect sexual behavior in two ways: they first organize the structure and function of brain cells and tissues during development, and then at puberty and in adulthood they activate these functions. That is, the circuits that support sexual behavior must be organized and then, at some later time, activated.

Organizational effects have the following characteristics: they occur only during a "critical period" early in development, usually during gestation or just after birth, but not later in life. They are permanent, persisting throughout life, and are opposite in males and females. These characteristics imply structural changes or other long-term cellular changes in the brain.

In contrast, activational effects generally occur at puberty and in adulthood, do not have a "critical period," are not permanent, and are similar in males and females. They are thought to involve subtle changes in brain function.

Although the development of sexual behaviors in most vertebrates, including mammals conforms to the organizational-activational hypothesis, many exceptions have been noted (Arnold & Breedlove, 1985). Nonetheless, it is useful in conceptualizing the effects of steroid hormones on the development of adult sexual behavior.

Development of Male Sexual Behavior

Observations in mammals including humans provide for the following general picture of how steroid hormones influence the development of the brain circuits that support male sexual behavior. The area of the brain thought to regulate sexual behaviors is the hypothalamus and areas of the limbic system.

As described in Chapter 3, genes on the Y chromosome are expressed very early in gestation. One of these genes, SRY, also known as testis-determining factor, directs the undifferentiated gonads to become testes. Soon thereafter, the testes produce testosterone (T), which circulates throughout the body. In neurons and non-neuronal cells of the hypothalamus and limbic system, T has two possible modes of action. It can bind to testosterone **receptors,** forming a T:R complex and regulate **gene expression** directly, or it can be taken up into target neurons and converted to the estrogen, 17 beta-estradiol (E), by aromatase. The resultant E then binds to its receptor, the beta-estrogen receptor (bER). Recent experiments in rats suggest that the aromatase conversion of testosterone to estrogen in the sexual **dimorphic** area of the hypothalamus affects partner preference in males (Paredes, Nakagawa, & Nakach, 1998; Houtsmuller et al., 1994). Within their respective target neurons, T:R or E:bER complex binds to unique areas of specific genes with four possible outcomes: (1) initiate gene expression, (2) block gene expression, (3) increase ongoing gene expression, or (4) reduce ongoing gene expression. The combined effects on gene expression provide for the organization of neuronal circuits and other structures that support adult male sexual behavior. We don't know how many genes are regulated in any given neuron or cell, but the range of behaviors observed in animals and humans suggests that the regulation of all the genes involved varies. Remember the kaleidoscope that we all played with at some time in our childhood? Every time you rotated the tube, a different pattern appeared. Complex behaviors are thought to depend on many genes, and the outcome of gene expression that supports complex behaviors is a bit like a kaleidoscope. The sexual behavior of each individual reflects the expression of a variable level of expression of a unique set of many genes. The gene expression outcome for each individual is like turning the tube on a kaleidoscope; the difference between individuals is like the difference in patterns produced by turning the tube just a little bit.

Development of Female Sexual Behavior

So, how does the development of sexual behavior proceed in a female fetus? The gonads, as with all developing tissues, are able to mark time with exquisite accuracy. If no testis-determining factor is present, the gonads wait until about week 13 and then develop into ovaries. Human fetal ovaries produce very low levels of both E and T; much less E than they will ultimately produce in adulthood, and about 1/10 to 1/20 the level of T produced by the fetal testes. Thus, the female fetal brain is exposed to only very low levels of E and T. However, neurons in the fetal female hypothalamus and limbic system are just as capable as those in the fetal male of binding, taking up, and utilizing T or converting T to E. The regulation of a similarly unique number of genes is thought to take place in the same way as in the male fetal brain, resulting in the organization of those neuronal circuits and structures that will support adult female sexual behavior.

Now we are ready to look at how changing the amount of available T and/or E influences brain development and adult sexual behavior.

STUDIES OF NORMAL ANIMALS

First let's look at some examples in normal, untreated animals where genital appearance and adult sexual behavior correlate with gestational exposure to testosterone. We will examine three in which a correlation exists between fetal exposure to elevated levels of testosterone and masculinization of neonatal appearance, juvenile sex-typical behavior, and adult sexual behavior. For an exploration of the diversity of animal sexual behavior see Bruce Bagemihl's *Biological Exuberance: Animal Homosexuality and Natural Diversity* (1999).

Freemartins

A well-known phenomenon in the animal husbandry of cows is "freemartins." Freemartins are the females of nonidentical (male-female) twins born to cows. Although the calf is a genetic female and appears female at early stages of gestation, at birth she frequently appears male; the genitals are masculinized, with an enlarged clitoris nearly the size of a penis. Freemartins often look like young bulls and as adults are often infertile. The explanation for how these females become freemartins involves testosterone. It is thought that testosterone released by the male twin's testes has access to the female twin by the exchange of blood due to fusion of the placental membranes (Hafez, 1993). It is concluded that the exposure of the female twin to testosterone during gestation drives the sexual and neurological development of the female twin toward the male pattern.

Two additional situations that have been examined scientifically are mice and other rodents and the spotted hyena.

Mouse Uterine Horn

For all mammals that give birth to litters of newborns, the fetuses are arranged in their own amniotic sacs with their own placentas in a structure called the uterine horn (Figure 4.1).

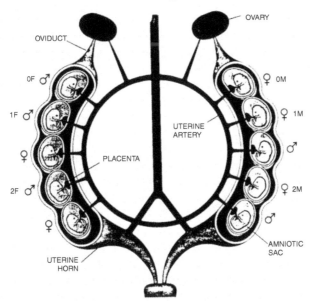

FIGURE 4.1. Rodent uterine horn: Hormonal environment in the uterus affects adult sexuality in mice. Female embryos surrounded by males on both sides (2M females) are exposed to higher levels of testosterone than females that do not develop next to a male (0M females). Mature 2M females have a masculine anatomy. They are also more aggressive, and are less attractive to males than are 0M females. The opposite, feminizing effect is seen in males surrounded by females (2F males). *Source:* Crews (1994).

The placement and number of male and female fetuses along the uterine horn is random and unpredictable. However, the positions of individual fetuses can be determined so it is possible to identify a female fetus that developed between two females (0M females, Figure 4.1) and one that developed between two males (2M females, Figure 4.1). So, we can ask whether a 2M female's external genitalia differ and whether its adult sexual behavior is comparable or different from that of a 0M female. A 2M female mouse has a longer anogenital space at birth and in adulthood relative to a 0M female. Data suggest that the anogenital space in female mice is a reliable predictor of the intrauterine androgen environment (Hotchkiss & Vandenbergh, 2005). Behaviorally, 2M females are more aggressive in a variety of test situations, they mark their territory with urine at a higher rate, they have longer and more irregular estrus cycles as adults, they are less attractive and sexually arousing to male mice, and they are less efficient at mating (vom Saal & Bronson, 1980). Last, they produce more male offspring than do 0M females (Ryan & Vandenbergh, 2002).

The fetal testes releases testosterone into the fetal blood supply, and the blood supply of each fetus has access to a common circulation system. Thus, we might expect the blood and amniotic fluids of a 2M female fetus to have higher levels of testosterone. In fact, fetal blood levels of testosterone were 50 percent greater than that of 0M female fetuses, but still less than that of normal male fetuses. Similar analysis of testosterone in 2M female fetus's amniotic fluid revealed a 25 percent increase relative to 0M female fetal levels. In contrast, blood analysis revealed that 2M adult blood levels of testosterone were not different from adult 0M female mice.

In rats, an area of the brain within the hypothalamus called the sexually dimorphic **nucleus** of the **preoptic area,** or SDN-POA, is larger in males than in females. Females with a longer anogenital space have been shown to have significantly larger SDN-POA than females with a small anogenital space (Faber & Hughes, 1992).

These results suggest that a correlation exists between the increased exposure of a female fetus to testosterone and male-like sexual characteristics of these animals as newborns and as adults. Thus, we can hypothesize that the normal variations seen in some sexual characteristics and behaviors of female mice are in part traceable to the female's variable exposure to testosterone during gestation, due to their intrauterine proximity to male fetuses.

If a female fetus positioned between two male fetuses displays more masculine sexual characteristics and behaviors, might a male fetus positioned between two male fetuses differ physically and behaviorally from a male positioned between two female fetuses? As we might expect, the answer is yes. The size and weight of the testes of male fetuses positioned between males are greater than that of males surrounded by female siblings. Their seminal vesicles, the receptacles at the base of the bladder and connected to the prostate gland that provide nutrients for and hold **semen,** are more sensitive to testosterone. Last, their aggressive behaviors are much more sensitive to stimulation by testosterone. These results demonstrate that increased exposure of males to testosterone during gestation increases the size of their genitals and the sensitivity in adulthood of both their genital tissues and brains to testosterone. In addition, both 2M females and 2M males receive more attention from their parents relative to their 0M sisters and 0M brothers.

Female Spotted Hyena

A second normal animal situation in which testosterone appears to masculinize females during gestation is *Crocuta crocuta,* the spotted hyena. Similar to most young mammals, a young spotted hyena (Figure 4.2) looks soft and cuddly. But be careful. Spotted hyenas are one of the most aggressive and ferocious mammals known. Let's see if we can understand why. Once again, we are simply observing behavior and analyzing testosterone levels and related enzyme activities in the blood and placenta.

The normal adult female spotted hyena is very different physically and behaviorally from other mammalian females. Her external genitalia are highly masculinized; the labia are fused to form a pseudoscrotum and the clitoris is enlarged to form a male-like phallus through which the urogenital sinus traverses. As a juvenile she engages in as much rough-and-tumble play as her male siblings. As an adult she is heavier and more aggressive than adult males, and is dominant in competitive situations with males. Within the matriarchal hyena society, social interactions between females include mounting and pelvic thrusting, which are male-typical behaviors. She

FIGURE 4.2. Female spotted hyena: Six-week-old spotted hyena. *Source:* Cover photo, *Science 260* (5116), 1993. Reprinted with permission. Copyright 1993 American Association for the Advancement of Science.

also marks her territory with urine, another male-typical behavior. Finally, during estrus, she dominates the male and determines the time of mating. The only other mammalian female that exerts similar control over mating is *Homo sapiens.*

Does a correlation exist between the adult appearance and behavior of the female spotted hyena and her exposure to testosterone during gestation? Testosterone and estrogen are derived from the precursor molecule androstenedione. The concentration of androstenedione in the blood of adult females is higher than in adult males, but the levels of testosterone are lower in adult females than in adult males. However, during pregnancy, the ovaries of the spotted hyena secretes substantial quantities of androstenedion, which is converted to testosterone and **estradiol** by the placenta (Lindeque, Skinner, & Miller, 1986; Glickman, Frank, Davidson, Smith, & Siiteri, 1987). Thus, the level of testosterone in the female increases and finally exceeds that in adult males (Yalcinkaya et al., 1993). Furthermore, the placenta of the spotted hyena contains 20 times more of the enzyme that converts androstenedione to testosterone than does a human placenta. Once again, a correlation exists between the masculinization of the female spotted hyena, both physically and behaviorally, with the increased levels of the placental enzyme that produces testosterone and the high circulating levels of testosterone in pregnant females. These observations support the hypothesis that the masculinization of the female hyena is due to the exposure of the female fetus to increased levels of testosterone during gestation.

Sheep

Within domestic sheep in the United States, approximately 8 to 10 percent of rams spontaneously display a stable sexual partner preference for other males and are thus classified as male-oriented rams. These rams are sexually inactive with **estrous** ewes. Studies have failed to identify any social factors that can predict or explain the homosexual behavior of these rams. All other aspects of their sexual behavior are typically male, with the exception of partner preference (for a recent review see Roselli, Larkin, Schrunk, & Stormshak, 2004). Male-oriented rams have also been observed in all-male groups of wild mountain sheep. Thus the homosexual orientation of these rams is not likely to be an artifact of human management, but rather a natural variant of the sociosexual interactions of sheep. Studies also demonstrate that same-sex mounting is not related to dominance rank within a herd or competitive ability.

Serum concentrations of testosterone do not differ between female- and male-oriented rams. Furthermore, castration reduces mounting behavior in both female- and male-oriented rams, but does not alter their choice of sexual partner. As in other mammalian males, prenatal testosterone is metabolized to estradiol by aromatase in the brain of rams. However, studies have shown that anatomical and functional differences exist in the brains of male-oriented versus female-oriented rams (Roselli, Larkin, Schrunk, et al., 2004). These differences are primarily within and functionally associated with the hypothalamus.

The **medial preoptic area** of the anterior hypothalamus is instrumental in mediating sexual motivation and partner preference (Paredes & Baum, 1997). Upon examination, aromatase activity in the preoptic area was found to be significantly lower in male-oriented rams than in female-oriented rams. This data and that primarily from other mammalian species suggests the hypothesis that male-oriented rams experience insufficient exposure to androgen-derived estrogens during the critical period in sexual development for sexual partner preference determination (Roselli, Larkin, Schrunk, et al., 2004).

Further studies of the **medial** preoptic area of the anterior hypothalamus by Roselli and collaborators has revealed a cell group referred to as the ovine **sexually dimorphic nucleus** of the medial preoptic area–anterior hypothalamus (oSDN) (Roselli, Larkin, Resko, Stellflug, & Stormshak, 2004). The volume of the oSDN is larger in males than females and is twice as large in female-oriented rams than in male-oriented rams. The number of neurons in the oSDN was greatest in female-oriented rams, followed in number by male-oriented rams, with the oSDN of ewes containing the fewest number of neurons. Further studies showed that neurons in the oSDN of female-oriented rams contain more aromatase mRNA than male-oriented rams, demonstrating that the oSDN neurons of female-oriented rams are more capable of converting testosterone to estrogen. It is suggested that these studies support the idea that a structure within the medial preoptic area–anterior hypothalamus influences sexual orientation in mammalian males (Roselli, Larkin, Resko, et al., 2004). I have presented the studies on sheep in some detail, because as we will see (Chapter 8) there are parallels to what we understand about the human hypothalamus.

ANIMAL EXPERIMENTS

Studies on the role of hormones in sexual development have primarily been done on rats. The general procedure for many of these experiments is to alter the availability of a steroid hormone to the developing fetus and then monitor the animal's adult sexual behavior. One of the early observations about sexual development that has helped with these studies is that the effects of steroid hormones on sexuality is restricted to what is called the "critical period." The critical period for sexual **differentiation** is the developmental time during which the steroid hormone can exert its effect. If the hormone is present before or after the critical period it will have no effect. It must be present during the critical period to be effective. In humans, the critical period is rather long and occurs only during gestation. However, in the rat, the critical period for neuro-organization and the establishment of sex-typical behaviors is from about embryonic day 16 through postnatal day 10. Since the gestational period for the rat is 20 to 21 days, this gives us a window of about 15 days during which we can experimentally manipulate hormone levels. After the hormone levels have been altered, the animal is allowed to mature, during which time its pre-pubertal behavior can be monitored. Finally, adult sexual behavior is monitored for differences from control animals of the same chromosomal sex. Control animals are subjected to the same physical procedures, but hormone levels are not altered.

Rat Mating Behavior

The adult sexual behavior monitored in females is **lordosis,** or the receptive posture assumed by the female to permit mounting by a male. The male typical behavior is mounting a receptive female and grasping of the female flanks with the front paws. Figure 4.3 shows the corresponding postures for females and males.

So, let's ask some simple questions about the effects of testosterone on the development of these sex-typical behaviors in rats.

Females Exposed to Testosterone

First, what happens if genetic females, animals possessing XX sex chromosomes, are exposed to high doses of testosterone during gestation? This is simply done by injecting a pregnant rat with testosterone. We want to observe the general physical appearance of the animal's external genitalia first and then the sexual behavior of these animals as adults. It is important to realize that for most mammals ex-

Hormonal Exposure and Mating Behavior in Rats

Mating behavior of rats is affected by exposure to hormones before birth. Males that receive insufficient androgens display stereotypically female postures, whereas females that receive an excess engage in stereotypically male behaviors. Extrapolating such data to sexual orientation, however, is difficult at best.

MALE MOUNTS FEMALE Male rat is considered heterosexual Female rat is considered heterosexual	FEMALE MOUNTS FEMALE Top female rat is considered homosexual Bottom female rat is considered heterosexual
MALE MOUNTS MALE Top male rat is considered heterosexual Bottom male rat is considered homosexual	FEMALE MOUNTS MALE Female rat would be considered homosexual Male rate would be considered homosexual (Not studied in experiments)

FIGURE 4.3. Rat mating behavior: The variations in rat mating behavior produced by varying exposures of fetuses before birth to androgen hormones. *Source:* Byne (1994).

cept humans, dolphins, Bonobo chimpanzees, and hyenas, sexual intercourse occurs only when the female ovulates or comes into estrus, or heat. It is only during estrus that a female can conceive, and often the only time she willingly participates in mating. When a female comes into estrus, males will persistently attempt to mate with her.

What does a female exposed to testosterone look like? Testosterone drives the development of the external and internal genitalia to the male form. In the absence of a Y chromosome, the internal female genitalia also develop. So, her external genitalia are indistinguishable from those of normal males, but she has both male and female internal genitalia.

Behaviorally, major changes are observed as well. Her capacity for the female receptive posture, lordosis, is greatly reduced compared to control females. That is, she does not posture receptively when males attempt to mount her. She displays much more mounting, or male-typical behavior, than control females. It is important to note that mounting behavior is not unique to males. Normal females do engage in mounting, but much less frequently than males. Nonetheless, mounting behavior by adult females treated with testosterone during gestation increases relative to control females. Upon treatment with male-levels of testosterone as adults, these females display a level of mounting behavior comparable to that of normal males and their capacity for the female behavior of lordosis is further suppressed. Last, she is more aggressive toward males than control females.

We can conclude that treatment of a female rat fetus with the male hormone testosterone results in an adult female that looks like a male and behaves more male than female. Furthermore, if she is treated with testosterone at a level equal to that circulating in adult males, her sexual behavior is comparable to adult males. Testosterone treatment during gestation and in adulthood essentially makes a female rat look and behave like a male rat.

Male Deprived of Testosterone

Now let's do an experiment where we take testosterone away from male rats during the critical period and see what happens to their adult behavior. Once again these are genetic males possessing XY sex chromosomes. The way this experiment is done is to remove the testes after birth, about one-third of the way through the critical period.

In this case we can assess only adult sexual behavior. These males develop the behavioral characteristics of genetic females. When presented with a female in estrus, their mounting frequency is lower than for control males. When injected with estrogen and progesterone to mimic the onset of estrus, they become sexually attractive to normal males and display the female receptive posture, lordosis, when mounted by a normal male. Not surprisingly, they are less combative than control males as adults. Consistent with the definition of the critical period, if the testes are removed after the critical period (in this case after postnatal day 10) the adults show little or no tendency to display lordosis under comparable conditions.

Female physiological functions can be observed in these males as well. **Ovulation** is controlled by the brain, specifically by the hypothalamus. Protein factors are released by the hypothalamus into the blood stream and eventually cause the ovaries to ovulate. If adult males denied testosterone at birth receive transplanted ovaries, these ovaries exhibit cyclic ovulation and the males display behavioral estrus. This result suggests that those brain centers that regulate ovulation are retained in males denied testosterone as newborns, and that in normal males the exposure of these brain centers to testosterone during the critical period eliminates this female brain function.

Additional Observations

A number of additional observations have been made in related animal experiments. These same hormone manipulations in both males and females carried out after the critical period do not affect the normal development and expression of sex-typical behaviors.

Treatment of normal genetic females (XX chromosomes) with a single dose of androgen (a steroid hormone related to testosterone) four days after birth, but within the critical period, permanently abolishes her ability to ovulate. Remembering that the brain regulates ovulation, this suggests that the function of those brain centers that control ovulation can be eliminated by this treatment during the critical period.

The nonreproductive behavioral activities of treated prepubertal juveniles are similarly affected. That is, males denied testosterone don't engage in as much rough-and-tumble play and females treated with testosterone don't engage in as much nest building relative to untreated control animals.

With regard to primates, female monkeys (XX chromosomes) treated with androgen during the critical period display more rough-and-tumble play, engage in more aggressive encounters with normal males, are less maternal imitative, and spend more time playing with other monkeys who were similarly exposed (regardless of their genetic sex) relative to untreated controls. So, an androgen-treated female monkey displays male sex-typical behaviors or the androgen treatment has masculinized female sex-typical behaviors.

The outcome for males is a bit more complicated. Genetic males (XY chromosomes) denied testosterone at birth and then treated with the androgen androstenedione (the end product of testosterone metabolism) display both male and female behavior as an adult. This result suggests that testosterone must effect both the masculinization and defeminization of selected brain centers. Other studies have demonstrated that some or all of the defeminization of the male fetal brain is dependent on the conversion of testosterone to estrogen by the enzyme aromatase. In rodents, adult male mounting behavior can be reduced by inhibiting aromatase activity during the critical period. However, aromatase inhibition did not alter partner preference. A recent similar experiment has been conducted on sheep (Roselli, Schrunk, Stadelman, Resko, & Stormshak, 2006). Ten pregnant ewes were treated with the aromatase inhibitor ADT (1,4,6-androstatriene-3,17-dione) at a dose known to inhibit 85 percent of the aromatase activity in the fetal hypothalamus. The seven males born to the treated ewes were then tested for their male-typical sexual behavior and partner preference. While the rams treated with ADT prenatally showed significantly lower female-directed mounts, there was no difference in sexual behaviors between ADT treated and control rams. Furthermore, prenatal ADT treatment did not significantly alter partner preference (Roselli et al., 2006). So, at present, just what role aromatase may play in the determination of partner preference is still unknown.

In summary then, testosterone is required for both the physical and neurological development of males. The neurological development of males includes both the masculinization and defeminization of the brain. Furthermore, the absence of testosterone during the critical period results in the development of a male fetus (XY chromosomes) along female lines, both physically and behaviorally. Last, exposure of a female fetus (XX chromosomes) to testosterone during the critical period leads to the development of the fetus along male lines, both physically and behaviorally. Clearly, testosterone is one of, if not *the* critical factor, in mammalian sexual development. Furthermore, we must be open to the likelihood that sexually dimorphic behaviors may have unique hormone requirements with regard to which hormone—testosterone or estrogen, how much hormone, and when and how long the exposure during gestation.

Although the studies examined so far tend to suggest that male behavior and female behavior exist, we need to keep in mind that significant variability exists in what we call male and female behavior, particularly in humans. Furthermore, a considerable overlap of male and female behaviors exist, especially again in humans.

NATURE'S HUMAN EXPERIMENTS

We know in some situations natural variations in the availability of testosterone lead to changes in the physical appearance and behavior of adult rodents and the female spotted hyena. We also know that if we alter the availability of testosterone during development, the sexual behavior of adult rodents is altered. How do we interpret these results with respect to the development of human sexuality? Would the same outcome result if the availability of testosterone were altered in humans? As it turns out, there are human conditions in which the ability of testosterone to function is either partially or completely blocked, or the availability of testosterone and other androgens is limited. Let's take a look at these situations and ask the same question: If the function or availability of testosterone during gestation in humans is altered, do correlative changes occur in the individual's physical appearance and adult sexual behavior?

Androgen Insensitivity

The human condition in which a complete failure of all tissues to respond to testosterone and dihydrotestosterone is called androgen insensitivity syndrome, also referred to as complete androgen insensitivity or CAIS (Diamond, 1992). The incidence of this condition is about 1 in 20,000 live births, or about 0.005 percent (Thompson, McInnes, & Willard, 1991), and is a consequence of a gene **mutation** that alters the receptor for testosterone in such a way that the receptor no longer binds testosterone. The receptor for testosterone is a large protein that normally binds testosterone, forming a testosterone-receptor complex. This complex localizes to the nucleus of responsive cells and there acts as a transcription factor, regulating the expression of selected genes (Figure 4.4). In general, in all cells with androgen receptors, the expression of selected genes is altered (some genes are

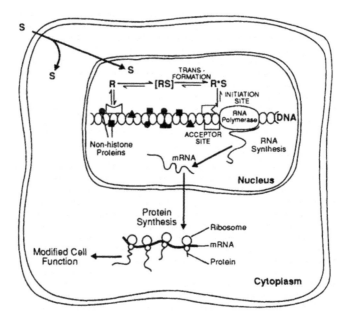

FIGURE 4.4. Steroid hormone-receptor complex regulation of gene expression: The nuclear model of the steroid hormone receptor. The steroid hormones (S) are distributed within both the cytoplasm and the nucleus of the target cell. The unoccupied receptors (R) are believed to be primarily concentrated in the nucleus in a reversible equilibrium binding state with the nonhistone proteins of the chromatin. Binding of the steroid hormone to its receptor results in the transformation of the unoccupied receptor (R) into the biologically active hormone-receptor complex (R*S) at the acceptor site. The transformed receptor then opens and initiation site on the DNA, resulting in the induction (or repression) of mRNA synthesis through the action of the enzyme RNA polymerase. The mRNA then enters the cytoplasm, where protein synthesis occurs on the ribosomes, resulting in modified cell function. *Source:* Brown (1994). Reprinted with the permission of Cambridge University Press.

turned on, some are turned off, and, for some, expression is increased or decreased) to provide for male functions.

Let's examine the consequences of androgen insensitivity on a human male fetus (XY sex chromosomes). The fetus has testes, so normal levels of testosterone are present, but neither testosterone nor dihydrotestosterone bind to their receptors. As a consequence, the internal and external male genitalia fail to develop. Müllerian duct inhibiting hormone is produced, so the internal female genitalia are eliminated normally.

A partial vagina develops and the testes are concealed, usually in the body wall or groin. In addition, none of the neuronal circuits in the brain that would ultimately control typical male sexual behavior are masculinized and/or defeminized. At birth the external genitalia of these individuals appear female, and as a result these infants are raised as girls. They develop a female outward appearance, but they fail to begin menstruation at puberty. Upon examination and testing, the defective androgen receptor is detected. So, what happens to these individuals? More to the point, what is their adult sexual behavior? They look and behave like typical females (Hines, 2000). Breast development is supported by estrogen originating from the testes and the aromatization of androstenedione and testosterone by aromatase. Pubic auxiliary hair is scant or absent. They have a female gender identity and are sexually oriented toward men. In two studies sexual attraction to men was reported by 100 percent of CAIS women during adolescence and 93 percent during adulthood (Money, Schwartz, & Lewis, 1984; Wisniewski et al., 2000). A study by Hines, Ahmed, & Hughes (2003) examined a number of psychological outcomes in 22 women with CAIS and 22 controls matched for age, race, and sex of rearing. Outcome measures included quality of life, including self-esteem and general psychological well-being, gender-related psychological characteristics, including gender identity, sexual orientation, and gender role behavior in childhood and adulthood, marital status, personality traits that show sex differences, and hand preference. No statistically significant differences were found between the 22 women with CAIS and the matched controls for any psychological outcome. The authors suggest that these finding argue against the need for two X chromosomes or ovaries for feminine-typical psychological development in humans. Furthermore, the data support the essential role of the androgen receptor for masculine-typical psychological development. Individuals with CAIS are usually surgically altered as needed to permit them to engage in heterosexual intercourse and are given estrogen as needed to enhance their female secondary sexual characteristics. They live their lives as females, marry, and raise adopted children. Long-term studies of these individuals have noted a striking absence of most male-typical behaviors and interests. Two examples of adult individuals with complete androgen insensitivity syndrome are shown in Figures 4.5 and 4.6.

So, the outcome for humans is similar to that observed for rodents: male fetuses deprived of the effects of testosterone during gestation develop an outward appearance of females and as adults display female sexual behaviors, including a female heterosexual orientation toward men.

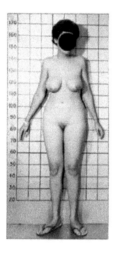 

FIGURES 4.5 and 4.6. Genetic XY males with complete androgen insensitivity: Complete androgen insensitivity syndrome (testicular feminization) in genetic males. Note female body contours, absence of underarm hair, sparse pubic hair, and breast development. This female appearance developed under the influence of the normal secretion of estrogens by the testes, with no additional hormone treatment. *Source:* Figure 4.5, Money & Ehrhardt (1972). Money, John and Anke. A. Ehrhardt. Man and Woman, Boy and Girl: Differentiation and Dimorphism of Gender Identity from Conception to Maturity. pp. 116, Fig. 6.4. © 1973. Reprinted with permission of The Johns Hopkins University Press. Figure 4.6, Thompson, McInnes, & Willard (1991). Reprinted from *Thompson & Thompson Genetics in Medicine,* M.W. Thompson, R.R. McInnes, and H.F. Willard, copyright 1991, with permission from Elsevier.

5-Alpha-Reductase Deficiency

A second condition that also affects human males is 5-alpha-reductase deficiency or 5-alpha-RD (Diamond, 1992). As mentioned previously, the development of the male external genitalia, the penis and scrotum, require a derivative of testosterone, namely dihydrotestosterone (DHT), which is produced from testosterone by the enzyme 5-alpha-reductase (Grumbach & Conte, 1998). There is a gene mutation that results in an inactive form of the 5-alpha-reductase, with the consequence of no DHT being available to provide for penis and scrotum development. However, the testes are present, once again usually concealed in the labia or groin, and testosterone is available, so male development of the internal genitalia and the brain does occur normally. At birth, males with 5-alpha-reductase deficiency appear female for the most part, although there are exceptions. As a result, these individuals are raised as girls. However, at puberty, their world is turned upside down.

As these outwardly appearing females reach puberty their concealed testes produce the male characteristic surge of testosterone. In short order, these girls take on the appearance of young men; they undergo male muscle development, their voices become more male-like, a scrotum develops, and what was considered at birth to be a clitoris grows into a penis capable of erections and ejaculation.

So, what happens to these folks? How do they and their families adjust to this inversion of sexual appearance and the implied change in their expected sexual and sex-typical behaviors? What is the effect of this inversion with regard to sexual orientation? This condition is comparatively rare, but two groups have been studied, one in New Guinea and one in the Dominican Republic. The ease with which these individuals make the transition from being girls to being young men appears to depend in large measure on how rigid the mores and moral attitudes of their community are.

In the Dominican Republic community, although the transition was not trivial for the individuals with 5-alpha-RD, the community was generally tolerant and accepting of the individuals with their new sexual identity. In one group of 18 individuals who had been raised as girls, 17 gradually accepted their identities as men. Eventually, 15 of them, who self-identified as heterosexual, married; some raised adopted children. These individuals assumed traditional male gender roles, including working at a "typical man's job" usually involving heavy labor. One individual of this group would be classified as homosexual or possibly transgendered, as he chose to maintain his identity as a woman.

An additional observation regarding environmental influences on sexual orientation can be made from this study. From birth onward these individuals were raised as girls presumably within families with heterosexual parents, where typical heterosexual male and female behaviors predominated. Their familial and cultural environment gave them clear examples of what behavior was acceptable for them and expected of them as girls. Nonetheless, once they physically changed from girls to young men, the majority of them seemed to have no difficulty expressing a heterosexual orientation and male sex-typical behaviors. If familial and cultural environment can have an influence on sexual orientation, we would expect more of these individuals to express a homosexual orientation. The fact that only 1 of the 18 individuals could be classified as homosexual supports the conclusion that sexual orientation is not significantly influenced by family and/or culture. These results further support the theory that sexual orientation is determined during gestation and cannot be changed after birth.

The consequences for individuals with 5-alpha-RD in New Guinea are strikingly different. The Sambia are a warrior-based society that maintains rigid role divisions between the sexes. Women serve as beasts of burden, function as mothers and gardeners, and are regarded as dirty polluters. There are separate footpaths for men and women in their villages and separate living spaces for husband and wife within their huts. Boys leave their mother's care at the age of seven and are forcibly initiated into life in the communal men's house where boys and bachelors live as a group and women are banned.

In one study of 10 individuals with 5-alpha-RD, only 5 married and 1 committed suicide. If they marry while they still appear female, they are usually rejected by their husbands because they cannot engage in heterosexual intercourse since their vagina ends as a blind tube. In the end they often leave their homeland.

Another observation can be made with regard to New Guinea males. Part of the forced initiation of boys into life in the men's house involves performing fellatio or oral sex on the adult men. This practice starts at about age seven and continues to maturity. Once again, if sexual orientation were influenced by the cultural/social environment, one would expect such homosexual interactions to influence the sexual orientation of these men. This is not the case, however; the distribution of sexual orientation in these tribesmen is not significantly different than that found in other populations. Thus, as has been concluded by Imperato-McGinley (2002):

> it appears that the extent of androgen (i.e., testosterone) exposure of the brain in utero, during the early postnatal period, and at puberty, has more of an effect in determining male gender identity than does sex of rearing and sociocultural influences. (p. 124)

Congenital-Adrenal-Hyperplasia

A third human condition in which shifts in sexual orientation and sex-typical behavior appear to correlate with an increase in the availability of androgens during gestation is congenital-adrenal-hyperplasia (CAH). The most common form of CAH has an incidence of about 1 in 12,500 births or about 0.008 percent (Thompson et al., 1991). As the name implies, this is a condition that arises during development. The adrenal glands are located on top of the kidneys and are responsible for producing and releasing adrenal steroids and androgens. In the most common form of CAH, a genetically controlled enzyme deficiency causes abnormally high levels of androgen production, resulting in the availability of masculinizing levels of testosterone and DHT. These levels are in the normal male range during fetal development. Both males and females can acquire CAH, however in females it results in varying degrees of genital masculinization. In general, the genitals are surgically altered to appear female and the individuals are raised as girls.

The masculinization of the genitals during gestation implies that the brain was also exposed to masculinizing levels of androgens, so masculinization of sexual orientation and sex-typical behaviors would be expected. In fact, as juveniles, females with CAH display masculinized behavior patterns relative to unaffected females. They engage in more rough-and-tumble play, tend to prefer competitive sports, and display much less maternal imitative behaviors. As adults, they score better on average than other women on spatial tests, and there is a significant increase in the likelihood that they will be lesbian or bisexual.

In two separate studies of CAH women, 48 percent reported same-sex arousal imagery and 22 percent reported same-sex partner sexual contact. These percentages are three- and two-times greater respectively than the values reported by Kinsey (Kinsey, Pomeroy, Martin, & Gebhard, 1953). These data have been confirmed in two additional studies (Dittmann, Kappes, & Kappes, 1992; Zucker et al., 1996). In an additional study, a group of seven CAH women, whose genitals were extremely masculine at birth, were exclusively oriented toward women as sexual partners (Money, 2002). With regard to psycho-sexual development, a study of core gender identity, sexual orientation, and recalled childhood gender behavior found that women with CAH recall significantly more male-typical play behavior as children than did their unaffected female relatives (Hines, Brook, & Conway, 2004). They also reported significantly less satisfaction with the female sex of assignment and less heterosexual interest. Furthermore, among women with CAH, recalled male-typical play in childhood correlated with reduced satisfaction with female gender and reduced heterosexual interest in adulthood. Thus, the data suggest that those girls with CAH who show the greatest shift in childhood play behavior may be the most likely to develop a bisexual or homosexual orientation as adults and to be dissatisfied with the female sex of assignment.

Turner Syndrome

Turner syndrome affects only females and results from a lack of one of the two X chromosomes normally present in each cell. As a consequence, their genetic makeup is designated XO, and their ovaries fail to develop. While the development of the genitalia proceeds normally, in the absence of ovaries none of the normal gonadal hormones are available, including estrogen and testosterone. The only source of androgen in these individuals is the adrenal glands, however the adrenals are only a minor source of androgen during gestation. Thus, the brain develops under minimal masculinizing influence and remains maximally feminine. This is a rare condition with an incidence of 1 in 5,000 live female births or about 0.02 percent (Thompson et al., 1991), and has not been studied extensively.

So, what behavioral characteristics do these women display as children and adults? As young girls, they display a singular interest in dolls, playing house, and dress-up. They have no interest in competitive sports and are generally shy and retiring. As juveniles, they are intensely interested in romance, marriage, homemaking, and children. As adults, they score in the average range for women on verbal IQ tests, but display impaired visual-spatial and perceptual abilities. Their cognitive **phenotype** generally includes normal verbal function with relatively impaired visual-spatial ability, attention, working memory, and spatially dependent executive function. These deficits are considered related to the complex interactions between genetic abnormalities and hormonal, that is, estrogen and androgen, deficiencies (Ross, Roeltgen, & Zinn, 2006). As these women are not competitive and avoid confrontation, they would not make good doctors or lawyers. It can be inferred from studies of these women that the range of behaviors displayed by normal women is contributed to by some testosterone-mediated masculinization and/or defeminization of the female brain.

SUMMARY

The critical period for **sexual differentiation** is the developmental time during which steroid hormones can exert their effects. During the critical period, the brain is organized to provide for sexual behaviors. In the absence of testosterone or estrogen, the brain organization will support female behaviors. Testosterone is required to provide for male sexual behavior. Testosterone is modified in brain cells by enzymes. One enzyme converts testosterone to dihydrotestosterone and the other, aromatase, converts it to estrogen, primarily in neurons of the hypothalamus and limbic system. Both are required to provide for typical adult male behavior.

It is thought that during the critical period, testosterone, estrogen, and perhaps other steroid hormones organize the neuronal circuits that will provide for adult sexual behavior in both males and females. Once organized, these circuits remain inactive until puberty when the pubertal surge of primarily testosterone and estrogen activate the circuits. The steroid hormones act by forming a complex with their target receptors. The steroid-hormone receptor-complex acts as a gene transcription regulator with four possible outcomes for each of the genes that it regulates: (1) initiate gene expression, (2) block gene expression, (3) increase ongoing gene expression, or (4) decrease gene expression.

Observations in animals demonstrate that females can be physically and behaviorally masculinized by exposure to testosterone during the critical period of development. Those situations that lead to exposure of female fetuses to testosterone and masculinization are freemartins, or the female of twin calves, a female mouse fetus positioned between two male fetuses, and the female spotted hyena.

Recently, male-oriented male sheep have been studied. Approximately 10 percent of domestic male sheep display a stable sexual preference for other males; they are referred to as male-oriented rams. The medial preoptic area–anterior hypothalamus is instrumental in mediating sexual motivation and partner preference. Studies of the brains of male-oriented rams reveal that aromatase activity in the preoptic area is lower than in female-oriented rams. Furthermore, the ovine sexually dimorphic area of the medial preoptic area–anterior hypothalamus (oSDN) is twice as large in female-oriented rams than in male-oriented rams. These data suggest that a structure within the medial preoptic area–anterior hypothalamus influences sexual orientation in mammalian males.

Experiments in rodents have enabled us to define the critical period for their sexual differentiation. Experiments demonstrate that female fetuses exposed to testosterone are born appearing male and as adults act male when provided with adult male levels of testosterone. Similarly, males castrated at birth display female-like sexual behavior as adults. Additional observations have led to the following conclusions: Testosterone is required for both the physical and neurological development of males. The neurological development of males is thought to include both the masculinization and feminization of the brain. Furthermore, it is likely that sexually dimorphic behaviors have unique hormone requirements; that is, how much of which hormone, at what time, and for how long.

Complete androgen insensitivity is a human condition in which testosterone cannot function correctly throughout development and adult life. These individuals appear female at birth, and are raised as girls. They develop as psychosexual females; they see themselves as women, display female-typical behaviors, and are sexually oriented toward men.

Another condition in males is 5-alpha reductase deficiency. They have testicles that are concealed, and their external genitalia do not develop, so they appear female at birth and are raised as girls. The pubertal testosterone surge drives their physical development to the full adult male form. Even though they are raised as girls in presumably heterosexual households, they have a male gender identity, assume male-typical behaviors, and are sexually oriented toward women.

Congenital adrenal hyperplasia is a condition that results in masculinizing levels of androgen in the developing fetus. Both males and females can be affected, however in females it results in varying degrees of genital masculinization. The brain is also exposed to elevated levels of androgen. As juveniles these females display masculinized play behavior. As adults they score better than average than other women on spatial tests, and there is a significant increase in the likelihood that they will be lesbian or bisexual.

Turner syndrome is a chromosomal abnormality that affects females and results from the lack of one of the X chromosomes normally present in each cell. As a consequence, the ovaries fail to develop, and they are not exposed to the steroid hormones, including estrogen and testosterone, that they would be normally. The brain also develops in absence of these steroids. As juveniles and adults, they are extremely feminine. They display impaired visual-spatial ability, attention, working memory, and spatially dependent executive function. They are not competitive and avoid confrontation.

These human conditions mirror the results of animal observations and experiments. Males need testosterone to develop the physical and behavioral characteristics of typical men. Females appear to require little testosterone to develop the physical and behavioral characteristics of typical women.

Chapter 5

Gestational Neurohormonal Theory

"Sexual orientation in all mammals is primarily determined by the degree to which the nervous system is exposed to testosterone and certain other sex hormones while neuro-organization is taking place, and hormonal and neurological variables, operating during gestation are the main determinants of sexual orientation." Neuro-organization refers to the development of the brain and nervous system. This was the hypothesis put forward by Ellis and Ames in their 1987 article titled "Neurohormonal Functioning and Sexual Orientation: A Theory of Homosexuality-Heterosexuality" (p. 251). The article was the result of an extensive search of the scientific literature prior to 1987 concerning the biological basis of sexuality. They put forth a theory of human sexuality development referred to as the gestational neurohormonal theory, which I shall refer to as simply the neurohormonal theory. They also presented the following predictions regarding homosexuality based on their theory:

1. Homosexuals should have higher frequencies of sex-typical behaviors normally associated with the opposite sex.
2. Relationships between parents and homosexual offspring often may be strained and/or assume some cross-sex characteristics.
3. Homosexuality should reflect a significant degree of hereditability.
4. Average neurohormonal differences should exist between homosexuals and heterosexuals in both sexes at comparable ages.
5. Attempts to alter sexual orientation after birth should be minimally effective.
6. Homosexuality should be primarily a male phenomenon.

Nature's Choice

Before we look more closely at these predictions, let's clarify what we mean by *neuro-organization* and *sex-typical behavior.*

NEURO-ORGANIZATION

Neuro-organization is the entire process that gives rise to a functional brain and nervous system. Ellis and Ames (1987) define two stages of neuro-organization with regard to sexual orientation and sex-typical behavior. The first stage occurs roughly between the middle of the second and the end of the fifth month of gestation, and appears to determine sexual orientation. Figure 5.1 shows a human embryo at nine weeks of gestation. At eight weeks of gestation, it is nearly half this size. It is important to realize just how small an embryo is at this stage of development. Furthermore, because of the embryo's small size, the distances between brain structures is very

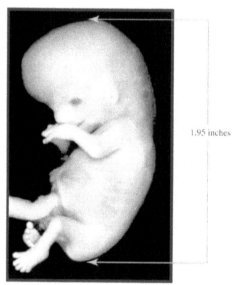

1.95 inches

FIGURE 5.1. Nine-week-old human fetus: The crown-to-rump distance is on average 50 mm, or 1.95 inches, and weighs on average 8 grams, or 0.28 ounces. An eight-week-old fetus has a crown-to-rump length of 30 mm, or 1.17 inches, and weighs on average 4.8 grams, or 0.17 ounces. *Source:* Carlson (1994). Reprinted from *Human Embryology and Developmental Biology,* B.M. Carlson, copyright 1994, with permission from Elsevier.

small, and small changes in the concentration of steroid hormones can have significant effects on development. It should also be noted that the cortex has not started to develop at this stage. So, those areas of the brain thought to be involved in sexual orientation determination develop well before the cortex or thinking brain develops. Thus, there is not likely to be a relationship between sexual orientation determination and our ability to think. The second stage of neuro-organization overlaps the latter part of the first stage (extending beyond it for at least two to three more months) and appears to determine a number of sex-typical behavior patterns.

SEX-TYPICAL BEHAVIOR

Sex-typical behavior is behavior that is more common or intense in one sex than in the other, regardless of cause. In children, for example, rough-and-tumble play is considered a male-typical behavior, while playing with dolls and playing house are considered female-typical behaviors. These and other sex-typical behavior patterns that have been found to be nearly universal in humans have also been documented in a wide variety of mammalian species. Furthermore, laboratory experiments in mammals (usually rodents) show that sex-typical behaviors can be inverted by manipulating neurohormonal factors, especially during brain development. For this reason, Ellis and Ames (1987) also hypothesized that many sex-typical behavior patterns in humans substantially reflect the effects of neurohormonal factors during gestation.

DEDUCTIONS

With these definitions of neuro-organization and sex-typical behavior in mind, let's look at the predictions put forth by Ellis and Ames. Their first prediction is that homosexuals should have higher frequencies of sex-typical behaviors normally associated with the opposite sex. This is generally true, as a large meta-analysis of studies has found that homosexuals report more opposite sex-typical or gender-nonconforming behavior as children than did heterosexuals. In fact, extreme opposite-gender sex-typical behavior has been shown to be predictive of homosexuality in both men and women (Bailey &

Zucker, 1995). A meta-analysis is a study conducted using data from a number of similar previous studies. It allows for the analysis of a large data set. Furthermore, other studies confirm that very feminine boys tend to become gay men and very masculine girls have a high likelihood of becoming lesbians (Skidmore, Linesenmeier, & Bailey, 2006). However, not all gay men are feminine and not all lesbians are masculine. Thus, a relationship appears to exist between gender-nonconforming behavior and homosexuality, but it is not completely understood. However, this relationship appears to be stable across different cultures.

The second prediction Ellis and Ames (1987) made is that homo-sexuality should reflect a significant degree of hereditability. We will discuss this later, but for now we can say that some evidence supports a genetic influence in both male and female homosexuality. At present, we do not have enough data to determine whether all male and female homosexuality is influenced or determined by genes. We must remain open to the likelihood that some fraction of homosexuality in both males and females occurs independent of a gene or genes.

The third prediction is that average neurohormonal differences should exist between homosexuals and heterosexuals in both sexes at comparable ages. Recall that the neurohormonal theory states that those processes that determine sexual orientation and sex-typical be-haviors occur during gestation. Therefore, differences in neurohor-monal levels that participate in these processes would have to be mea-sured as early as the second month through the eighth month of gestation. We can define this period from the second through the eighth month of gestation as the critical period for sexual orientation and sex-typical behavior dependence on neurohormonal factors in humans.

To date, no studies of hormone levels during the critical period in humans have been conducted, so we do not have definitive proof that there is a correlation (positive for lesbians and perhaps negative for gay men) between gestational testosterone levels and adult homosex-uality. However, experiments in animals do support this deduction. Recall from Chapter 4 that the critical period for rats ranges from embryonic day 16 through postnatal day 10. It has been shown that reducing the levels of the steroid hormone testosterone during this critical period alters sexual orientation behavior in male rats, as does increasing testosterone levels during the same period in female rats.

Ellis and Ames (1987) made two more predictions. The fourth is that attempts to alter sexual orientation after birth should be minimally effective, which is now generally accepted and supported by both the American Psychiatric Association (2000) and American Psychological Association (1997).

The fifth prediction is that homosexuality should be primarily a male phenomenon. This is borne out by the estimated levels of homosexuality in the general population, which are about 3 to 4 percent for males and about 1 to 2 percent for females. These levels of homosexuality were confirmed in a recent study of 8,000 American and Canadian college students, which found the proportion of homosexuals and bisexuals combined to be 3 percent of the male participants and 2 percent of the female participants (Ellis et al., 2005). This is about half the frequently cited level of 10 percent. This 10 percent level derives from the Kinsey reports (Kinsey, Pomeroy, & Martin, 1948; Kinsey et al., 1953), where it referred only to men and only those who had been predominantly homosexual for at least three years of their adult life. Nonetheless, homosexuality is about twice as prevalent in males as in females.

Let's ask one more question. Based on the neurohormonal theory, what would be the expected outcome of a study of a trait with respect to sexual orientation? How would we decide which trait to study? The best indicator of which trait might vary with respect to sexual orientation is that the trait is different between men and women. Such a trait would be classified as sexually dimorphic.

We would conduct this experiment by first identifying groups of heterosexual men, gay men, lesbians, and heterosexual women. We would try to make these groups as large as possible. Next we would obtain permission from the study subjects for their participation in the study. Then we would have them fill out questionnaires to assess their sexual orientation. In order to maximize the differences in the measured trait between groups, we would exclude those subjects that did not score at the extremes of the sexual orientation scales. In practice, this means that we would not want to include bisexual individuals in the study. Then we would measure the trait of interest. What would be the expected result based on the neurohormonal theory?

Let's say that the size of a brain area is what we want to measure, and that it could be measured noninvasively, that is, we could make the measurements on live subjects. Also, let's say that the brain area of interest is larger in men than in women. The neurohormonal theory would describe gay men's sexual orientation behavior as feminine relative to heterosexual men and lesbian sexual orientation behavior as masculine relative to heterosexual women. So, the anticipated rank order of brain area size would be heterosexual men larger than gay men, gay men larger than lesbians, and lesbians larger than heterosexual women. Graphically this could be presented as below.

I————I————————————I————I
Heterosexual   Lesbians   Gay   Heterosexual
    Women   Men   Men

Brain Area Size Increasing →

This is where our understanding of the biological basis of human sexuality stood in 1987. Next I will present a brief discussion of scientific data and its interpretation as a prelude to presenting data from studies of sexual orientation in humans.

SUMMARY

The neurohormonal theory states that sexual orientation in all mammals is primarily determined by the degree to which the brain is exposed to testosterone and other sex hormones while brain development is taking place. Furthermore, hormonal and neurological variables operating during gestation are the primary determinants of sexual orientation. The theory predicts that the result for women would be increasing levels of testosterone leading to lesbianism. For males, the initial interpretation would be that gay men develop due to reduced levels of testosterone. As we will see, the theory has been a guiding influence on the study of homosexuality in humans. With respect to women, the theory has been accurate as far as it goes. For men, the theory appears to account for only a fraction of the data and may not apply to a separate subfraction of gay men. Overall, it has been fairly accurate and has guided a great deal of work that has increased our understanding of the factors that contribute to homosexuality in humans. Furthermore, in doing so, it has also guided us in understanding the factors that contribute to heterosexuality as well.

It predicts that human sexuality is composed of sexual orientation and sex-typical behaviors. These two aspects of our sexuality develop during gestation, with the period for sexual orientation development preceding and overlapping the period for sex-typical behavior development. Both are subject to the influences of testosterone; testosterone can influence sexual orientation only, or it can influence both sexual orientation and sex-typical behaviors.

It predicts that homosexuals should have higher frequencies of sex-typical behaviors associated with the opposite sex, and this is generally true. It predicts that sexual orientation should reflect a significant degree of hereditability, and this is true.

It predicts that average neurohormonal differences should exist between homosexuals and heterosexuals in both sexes. This may be true, but there are no studies of hormone levels during the critical period for humans (second to eighth month of gestation). Thus, we do not have proof that a correlation exists between testosterone levels during gestation and adult homosexuality.

Attempts to alter sexual orientation are predicted to be minimally effective, and this is now generally accepted. Last, it is predicted that homosexuality should be primarily a male phenomena, and it is, as there are about twice as many gay men as lesbians (3 to 4 percent gay men versus 1 to 2 percent lesbians in the general population). It is an excellent first-order theory, and only time will tell how accurate its predictions will remain.

Chapter 6

Science

Before we look at studies of human sexual orientation, I will present a definition of science, some aspects of experimental design, and scientific data evaluation. Last, I will present an overview of how basic scientific research is funded in the United States.

WHAT IS SCIENCE?

First, let us be clear about what science is. Science is defined as the state of knowing; knowledge as distinguished from ignorance or misunderstanding. In *Shadows of Forgotten Ancestors,* Carl Sagan and Ann Druyan elaborate on this definition of science and put it in social context (Sagan & Druyan, 1992). They characterize science as follows:

> Science is never finished. It proceeds by successive approximations, edging closer to a complete and accurate understanding of Nature, but it is never fully there. (p. xiv)

> Science is always subject to debate, corrections, refinement, agonizing reappraisals, and revolutionary insights. (pp. xiv-xv)

> . . . . the cure for the misuse of science is not censorship, but a clearer explanation, more vigorous debate, and making science accessible to everyone. (p. 68)

Nature's Choice

EXPERIMENTAL DESIGN

In the biological sciences, a frequently encountered experimental design is the study of extremes. As an example, consider the study of what makes people tall. If you want to understand the processes and identify the molecules that contribute to height, you would use the experimental design of studying extremes; that is, you would study very tall versus very short people. The study of extremes maximizes the difference in the absolute value of characteristic determinations between two populations. The study of intermediate forms will not yield the desired maximum differences. With respect to human sexuality, identifying the processes and molecules that underlie the development of homosexuality provides an understanding of the processes and molecules that determine heterosexuality. Because bisexuality is viewed as an intermediate form, it has not been included in most of the studies reported thus far.

WHAT SCIENCE CAN TELL US

With this understanding of what science is, let's review some aspects of scientific data and its interpretation as well as its limitations. The studies we are going to examine will present data in the form of distributions, averages, and ranges. Referring to Figure 6.1, we define these terms and values as follows:

> *Distribution:* The position and arrangement of the data over all determinations.

Figure 6.1 shows an example of a normal distribution, which is also commonly referred to as a bell-shaped curve.

Measurements of many biological properties are distributed as a bell-shaped curve. Depending on the property being examined, the shape of the curve will vary from a tall and narrow to a short and wide curve.

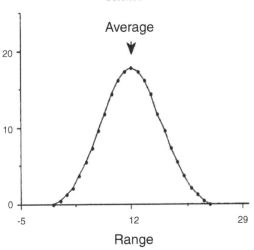

FIGURE 6.1. Normal distribution: Example of a normal distribution, often referred to as the "bell-shaped curve" showing the *range,* or the values between the lowest and highest value, and the *average,* or the mean, the sum of all the values divided by the number of determinations. The measurements of many things in nature distribute as a normal distribution. This is particularly true for size and timing, or how fast processes occur.

Average, arithmetic mean, or mean: An average is the sum of all the data points divided by the number of data points, and is shown in the figure for a bell-shaped curve as the value at the top of the curve.

Range: The range of a collection of values is simply the inclusive values from the lowest value to the highest value.

Now let's do a simple experiment to illustrate how these terms are used, and what the limitations are on the interpretation of data reported as means or averages. Suppose we have a room filled with one hundred people, fifty women and fifty men. Now, let's measure their heights and plot the data as shown in Figure 6.2, or in scientific terms, the number of men or women of a particular height is plotted as a function of their height. If we measured enough people, we would get a *distribution* as shown in Figure 6.2, once again a bell-shaped curve. We see that the curve for men overlaps the curve for women. In scientific terms, the ranges for the two curves overlap. However, they do

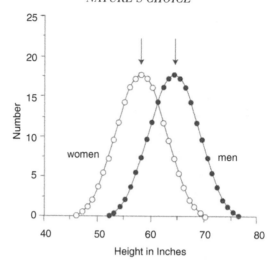

FIGURE 6.2. Example experiment: Data from a hypothetical experiment in which the heights of a large number of men and women are determined and plotted as a function of the number of individuals of that height. The average or mean values are indicated by arrows. The important point is that the ranges of the two groups overlap significantly. In most cases, knowing the height of an individual would not necessarily indicate whether the individual was a man or a woman.

have distinct averages as indicated by the arrows, and we can say that, on average, men are taller than women. So our experiment has yielded some useful information. We know the average height for men and for women as well as the height of the shortest man and woman and the tallest man and woman. But if I told you the height of an individual was 60 inches, could you tell me if the person was a man or a woman by looking at the data plotted in Figure 6.2? The answer is no. The range of the values for men and women overlap, and includes the stated value of 60 inches. Knowing the average height of men versus women does not allow us to say anything about the sex of a person of a particular height. Or, put in general terms, population statistics do not yield information about individuals. There is a limitation to how we can use the data. So, when someone tells you that the average of a certain measurement is different for men and women, realize that there is no predictive value in that difference unless the ranges for the measurements are also very different for men and women.

SCIENCE IN THE UNITED STATES

When presenting this material, I am often asked why more research is not done on human sexuality, specifically sexual orientation. So I include here a brief discussion that I hope helps you understand how science, particularly biological science, is funded in the United States. Scientific research in the United States is conducted by two general groups: (1) nonprofit organizations such as academic institutions and private not-for-profit institutes, and (2) for-profit organizations, primarily pharmaceutical and biotechnology companies.

Scientific studies conducted by pharmaceutical companies are primarily directed toward the discovery of drugs to treat specific conditions. They are not intended to make substantial contributions to our general knowledge about how the human body develops or functions.

Research at academic and not-for-profit institutions is conducted somewhat differently. I will describe how academic research is conducted, as it is what I know best. Universities and nonprofit institutions hire researchers who possess expertise in a specific area of science. They are not hired to work on specific projects. The academic institution does not provide the faculty researcher with funds to conduct research. The researcher must obtain funds for research by writing grants.

The largest source of funds for academic research in the biological sciences in the United States is a branch of the Public Health Service known as the National Institutes of Health, or NIH. The NIH awards grants for research at least loosely related to medical conditions or medicine. The National Science Foundation, or NSF, is also a source of federal research funds. The NSF funds research over all areas of scientific interest. No direct attempt has been made by NIH or NSF to control research conducted by academic researchers. These organizations do, however, solicit grants in areas that their advisors consider important. For example, when AIDS became a national health problem, the NIH issued requests for proposals (RFPs), that were directed at certain aspects of the AIDS problem.

All grant proposals are evaluated by review committees composed of academic researchers from around the country with demonstrated expertise in the area of the grant under consideration. Reviewers evaluate the proposals for, among other things, the importance of the research, the technical capability of the researcher, and the quality of support given to the researcher by their institution. Reviewed proposals are ranked relative to others reviewed at the same time. The NIH recommends funding of proposals, based on rank, which are ultimately approved and funded by Congress. Congress has the right to refuse to fund a grant even if it is recommended for funding by the review committee. Thus, politics plays a role in this process. If you are interested in reading more about this topic, go to your library and look for the *New York Times* editorial by Bob Herbert titled "The Big Chill at the Lab" (Herbert, 2003). If the subject of a grant is not of significant interest to the members of the review committee, it will generally not be funded. So, although academic researchers have considerable freedom to work in any area they want, they must work in an area that the general scientific community also considers important. It is a balancing act. You can do anything you like as long as it is fundable. In the end, a grant once funded is a contract between the researcher and his or her institution and the NIH. The researcher is expected to conduct the studies as described in the funded proposal and publish the results in an appropriate peer-reviewed professional journal.

SUMMARY

Science is the pursuit of knowledge. It is never finished. We are constantly moving closer to a complete understanding of nature, but we will never achieve that end. Nonetheless, we must try to make the knowledge we gain through scientific endeavors accessible to everyone.

One scientific method is the study of extremes. We design experiments to compare a trait between groups in an attempt to identify the molecules and processes that control the development of that trait. In such a process, the data we collect has a distribution, a range, and is often reported as the average or mean for each group. These averages or means tell us something about the population that each group represents, but they can't tell us anything about an individual member of the group.

Last, in response to the question of why more work isn't done on sexual orientation, it is really a matter of what individual researchers want to work on and what they think would get funded. Politics can play a role in the primary process for funding academic research in the United States.

Chapter 7

Genetics

Now let's take a look at what we know about the genetics of sexual orientation. Questions about the genetics of any human trait can be asked in two ways. One is by what is called population genetics, which means we examine the appearance of the trait across generations of individual families. A more rigorous population genetic technique is to determine the probability of a trait being found in both of monozygotic or identical twins, dizygotic or nonidentical twins (also just siblings), and adoptive siblings. The second method is genetic **linkage** analysis, which involves examining the **DNA** of family members to determine if a correlation exists between the appearance of the trait of interest and a unique sequence of DNA at a precise location on a specific chromosome. Finding such a correlation implies that the trait is carried by a gene within or containing the unique DNA sequence. First let us see what population genetics or studies of families can tell us about sexual orientation.

FAMILY STUDIES

Scientists who conducted the early population genetic studies of sexual orientation looked at familial traits. A familial trait is shared by family members or runs in families. The term *familiality* is defined as the more common appearance of a trait in the relatives of a positive individual than in the general population, whether the cause is genetic, environmental, or both. In two studies, homosexuality was determined to be familial for both males (Pillard & Weinrich, 1986) and females (Bailey & Benishay, 1993). That is, self-identified homosexual men had about four times as many homosexual brothers than did

Nature's Choice

heterosexual men, while self-identified lesbians had four to five times as many lesbian sisters as did heterosexual women. The median rate of homosexuality in brothers of gay men is about 9 percent, while that for sisters of lesbians varies from 6 percent to 25 percent (Bailey & Pillard, 1995). Thus, both male and female homosexuality run in families. But it is not clear whether both male and female homosexuality run in the same families.

A third study used population genetic techniques to look at familiality of male and female homosexuality (Bailey & Bell, 1993). The results again suggest that male and female homosexuality runs in families. Furthermore, this investigation did not find evidence for environmental influences on sexual orientation; neither male nor female sexual orientation appeared to depend on parental influences, suggesting that environment is not the dominant determinant of sexual orientation. Last, the study was not able to find any evidence for independent causes of male and female homosexuality, suggesting that the causes are similar or the same for both males and females. This is an interesting finding in light of the evidence that levels of testosterone appear to be involved in both male and female homosexuality. Once again these data are consistent with the neurohormonal theory.

The next study published was a population study that addressed the familiality of sexual orientation in 358 women (Pattatucci & Hamer, 1995). The results of this study indicated that female homosexuality did indeed cluster in families. Furthermore, elevated rates of homosexuality were found in four classes of lesbian relatives: sisters, daughters, nieces, and female cousins through a paternal uncle.

Familial patterns of male homosexuality were further assessed to determine the rate of homosexuality and bisexuality of brothers and sisters of gay men (Bailey et al., 1999). For two groups of gay men consisting of 350 and 167 subjects respectively, the average percentage of siblings that were homosexual was 8.5 percent for brothers and 3.3 percent for sisters. Thus, gay men have about 2.6 times as many gay brothers as lesbian sisters.

The same study also assessed maternal versus paternal transmission of male homosexuality (Bailey et al., 1999). Using the same subject population plus an additional 65 sibling pairs, mixed results were obtained. A slight, nonsignificant excess of maternal uncles (4.4 percent and 2.5 percent) compared to paternal uncles (3.1 percent and

2.1 percent) was observed for the two largest subject groups, while for the smaller group, more paternal uncles (6.8 percent) than maternal uncles (4.5 percent) was observed. Clearly, it is apparent that how subject groups are defined and recruited can influence the results obtained. We will see that this does occur when we examine linkage studies. Overall, no evidence that male homosexuality is influenced by an X-linked gene was found in this study. Nonetheless, the authors state that their data suggest that X-linked genes account for relatively few cases of male homosexuality.

TWIN STUDIES

Traditionally, the most powerful way to study heritability for a given behavior or trait is to examine monozygotic (identical) twins, dizygotic (nonidentical or paternal) twins, and adoptive siblings. The power of this approach derives from monozygotic twins having the same genes—their DNA is 100 percent identical. The DNA of dizygotic twins and siblings is 50 percent identical (half of their genes are identical), while the DNA of adoptive siblings is nonidentical (they share genes to the same degree as do individuals in the general population). Furthermore, twin studies can differentiate between genetic and environmental influences. Additional studies would be required to determine if the environmental influences were either prenatal or family and cultural. Two such studies have been reported that address sexual orientation in men (Bailey & Pillard, 1991) and women (Bailey, Pillard, Neale, & Agyei, 1993). The primary question posed in both of these studies is: If one twin or sibling is homosexual, what is the chance that the other twin or sibling will also be homosexual? The data for both men and women are shown in Table 7.1. The percent of homosexual males and females in the general population is included at the bottom of the table.

Although a number of interesting observations can be made from these data, some aspects of the data are problematic. First, the positive findings: The probability of a second twin being homosexual if one twin is homosexual is very similar for men and women; 52 percent for men and 48 percent for women. This can also be stated slightly differently as there is a 48 percent chance that the second twin will *not* be homosexual for men and 52 percent chance for women. Thus, there is roughly a 50/50 chance of both twins being ho-

TABLE 7.1. The Probability of Second Sibling Being Homosexual

Sibling Pair	Men, % Probability	Women, % Probability
Monozygotic twins	52	48
Dizygotic twins	22	16
Nontwin siblings	9.2	14
Adoptive siblings	11	6
General population	3 to 4	1 to 2

Source: Adapted from Bailey & Pillard (1991) and Bailey, Pillard, Neale, & Agyei (1993).

mosexual for both men and women. Comparing these values with those for dizygotic twins or nontwin siblings we see that the probabilities drop by a minimum of 2.4-fold for men and 3-fold for women; the more similar the genetic makeup, the higher the probability that both siblings will be homosexual. Heritability can be measured, and values greater than 0.5 indicate a significant genetic involvement in a trait. Calculations of heritability of sexual orientation under a variety of assumptions about volunteer bias and population trait frequency yields values of 0.31 to 0.74 for males and 0.5 or greater for females (Pillard & Bailey, 1998).

In addition, both male and female homosexual monozygotic twin pairs were also highly similar in their gender atypicality scores, further supporting a genetic basis for this trait (Pillard & Bailey, 1998). Last, the few available examples of monozygotic male twins raised apart show a level of similarity in sexual orientation and sex-typical behaviors similar to monzygotic twins raised together (Eckert, Bouchard, Bohlen, & Heston, 1986; Whitam, Diamond, & Martin, 1993). Together, these data support the conclusion that homosexuality and sex-typical behavior are heritable, or strongly influenced by genes.

Now, some of the problems: The percent probability that a second sibling will be homosexual for dizygotic twins was reported as 22 percent for men and 16 percent for women. These values should be comparable to those obtained for nontwin siblings. Remember that both dizygotic twins and nontwin siblings have 50 percent identical DNA or genes. The values reported in these studies for nontwin siblings are 9.2 percent for men and 14 percent for women. Although the values for women agree fairly well, the values for men are significantly different. How do we understand this discrepancy? Well, in fact, there is no good way to deal with the discrepancy. Should we discount the entire study because all of the data do not fit the expected pattern? I believe the answer is no. What good scientists do is what the authors of this study have done; that is, they acknowledge the discrepancy and encourage other workers in the field to replicate the study to see if the discrepancy holds up or a different result is obtained. Clearly, more studies need to be conducted.

An additional problem exists. The probability of the second adopted sibling being homosexual is reported as 11 percent for men. Because adoptive brothers have no more genes in common than do men in the general population, the probability should be the same as or at least very close to the value reported for homosexuality in the general male population (2 to 4 percent). For women the corresponding data are 6 percent probability for adoptive sisters and 1 to 2 percent for homosexuality in the general female population. The biggest problem with this comparison is the uncertainty of the values reported for homosexuality in the general population for both men and women. The reason for this has to do with how it is determined. In most cases it is determined by verbal or written surveys, and here the critical issues are what questions are asked, how the answers are evaluated, and how large a population is surveyed.

Having said this, if we accept for the moment that the values presented in Table 7.1 are the best estimates available, then there is a twofold discrepancy between the probabilities reported for adoptive siblings relative to the general population for both men and women. The goal of both studies was to assess the heritability of homosexuality—which means the degree to which homosexuality can be explained by additive genetic differences. This assessment was done by first having a computer determine the best model that fit the raw data over a broad range of model assumptions. Then the researchers conducted a statistical evaluation to determine whether the values for heritability were statistically significant. Heritability for male and female homosexuality remained statistically significant over all model assumptions, for example, where the rate of homosexuality in the general population was set at the lowest or highest estimate.

In addition to heritability, the researchers also looked at those features of the environment shared by siblings (that is, the family and cultural factors that tend to make siblings more similar to one another), in all cases they determined those factors were less than the proportion attributed to heritability or zero. In layman's terms, this means *no* environmental influence on sexual orientation was found. So we see once again that family and cultural environmental factors probably make little or no contribution to the determination of either male or female homosexuality. Since absence of evidence is not evidence of absence, this study cannot be taken to mean that there are absolutely no family or cultural environmental factors that influence the determination of sexual orientation. However, at present, no credible scientific data have been published that identify such factors. Furthermore, scientific reports of the efforts to identify environmental factors have thus far failed to do so.

More recent studies (Bailey, Dunne, & Martin, 2000; Kirk, Bailey, Dunne, & Martin, 2000; Kendler, Thornton, Gilman, & Kessler, 2000) have used more sophisticated quantitative statistical analyses to assess environmental influences on sexual orientation. The first two studies also assessed covariation of childhood gender nonconformity (CGN) and continuous gender identity (CGI). Taken together, these more recent studies suggest the existence of both genetic and nonshared environmental influences on sexual orientation, CGN, and CGI. Nonshared environmental influences are factors that tend to make siblings different from one another.

Last, no significant support for the influence of shared environmental factors has been found, and most data indicate that genes influence sexual orientation more strongly in males than in females (Mustanski, Chivers, & Bailey, 2002). Overall, the population genetic data are consistent with and support the neurohormonal theory insofar as they demonstrate that sexual orientation is influenced by genes.

GENETIC LINKAGE ANALYSIS

The influence of genes on male sexual orientation has also been examined using molecular genetic techniques. In 1993, a group led by Dean Hamer performed the first study of sexual orientation at the molecular genetic level (Hamer, Hu, Magnuson, Hu, & Pattatucci,

1993b). The study consisted of two parts. First, the scientists identified 114 families that contained at least two male siblings, one of whom identified himself as homosexual. Examination of the extended family trees revealed that in some families more gay relatives were observed on the maternal side than on the paternal side (Figure 7.1). That is, the researchers found homosexuality to be significantly more common among maternal uncles than among males in the general population. This finding suggested that at least for some male homosexuals, the trait is inherited from their mothers and carried by female members of their family. Knowing this makes the job of looking for a gene that contributes to homosexuality in these males a little easier because it would be located on the X chromosome, which males exclusively inherit from their mothers. Both males and females inherit an X chromosome that is a cut-and-paste version of their mother's two X chromosomes. Hamer's group identified 40 families (about 35 percent of their original study group of 114 families) with two gay sons, but no gay father–gay son pairs for further analysis.

The second part of molecular genetic studies is linkage analysis. Linkage studies look for specific chromosomal regions that are passed down in families, along with a phenotype (in this case sexual orientation) at probability levels greater than chance or 50 percent for siblings. Hamer's group prepared and compared DNA samples from the 40 pairs of homosexual brothers and their available mothers and sisters with a series of 22 markers (unique DNA sequences) that spanned the X chromosome. The experimental protocol is shown in Figure 7.2.

By chance, both brothers would be expected to share or score positive for most markers 50 percent of the time. However, if a marker identified an area of the X chromosome containing a gene that influenced male homosexuality, then the chance that both brothers' DNA would be positive for that marker would be expected to exceed 50 percent. In fact, 33 of the 40 pairs of gay brothers, or 82.5 percent shared five markers associated with the tip of the long arm of the X chromosome designated **Xq28** (see Appendix). Applying technical corrections leads to a value of 64 percent. Seven of the 40 pairs of gay brothers, or 17.5 percent did not share these markers. The data are shown in Figure 7.3. This result suggests that the chromosomal region Xq28 contains a gene or genes that influences sexual orientation in some men. This does not mean that a "gay gene" has been localized

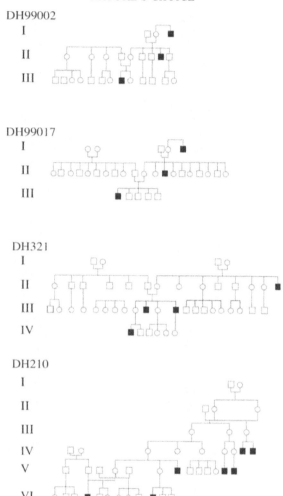

FIGURE 7.1. Family pedigrees: Family pedigrees displaying apparent maternal transmission of male homosexuality. Families DH99002 and DH99017 were from the randomly ascertained group of gay men. Families DH321 and DH210 were selected because many members were homosexual. Closed squares are homosexual males, open squares are nonhomosexual males, and open circles are nonhomosexual females. *Source:* Adapted from Hamer et al. (1993b). Adapted with permission from Hamer, D.H., Hu, S., Magnuson, V., Hu, N., & Pattatucci, A.M.L. (1993). A linkage between DNA markers on the X chromosome and male sexual orientation. *Science, 261*(5119), 321-327. Copyright 1993 American Association for the Advancement of Science.

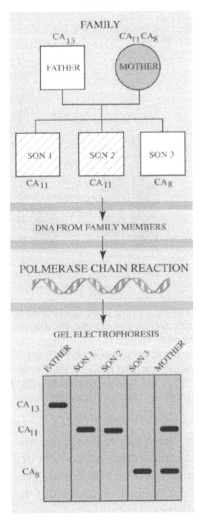

FIGURE 7.2. Experimental protocol: Pinpointing genes shared by gay brothers (striped boxes) first involved taking DNA from subjects. Several billion copies of specific regions of the X chromosome were then made using the polymerase chain reaction, and the different fragments separated by gel electrophoresis. Gay brothers shared a marker, in this hypothetical example, CA_{11}, in the Xq28 region at rates far greater than predicted by chance. *Source:* Le Vay & Hamer (1994). From "Evidence for a Biological Influence in Male Homosexuality" by Simon Le Vay and Dean H. Hamer. Copyright © 1994 by Scientific American Inc. All rights reserved.

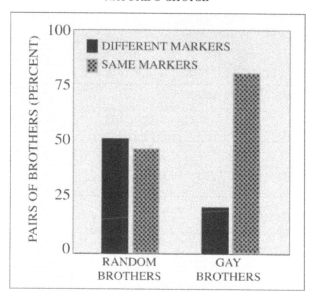

FIGURE 7.3. Hamer et al. first study data: Gene sharing in the Xq28 region is significantly greater in gay brothers than in the general population. Of 40 pairs of gay brothers studies, 33 pairs shared the Xq28 region. In a control group of 314 randomly selected pairs of brothers, the Xq28 markers were found to be almost equally distributed. *Source:* Adapted from Le Vay & Hamer (1994). From "Evidence for a Biological Influence in Male Homosexuality" by Simon Le Vay and Dean H. Hamer. Copyright © 1994 by Scientific American Inc. All rights reserved.

or identified. Within the Xq28 region there are about four million base pairs, representing less than 0.2 percent of the total human **genome** and likely containing several hundred genes. A great amount of work still needs to be done to confirm this result and eventually identify the gene or genes.

Shortly after this study was published, a letter and a technical comment were published. The authors pointed out weaknesses and technical inconsistencies that they believe make the results reported by Hamer's group questionable (Fausto-Sterling & Balaban, 1993; Risch, Squires-Wheeler, & Keats, 1993). As is characteristically done with such critiques, Hamer's group was invited to respond to the critique and both the critique and response published together (Hamer, Hu, Magnuson, Hu, & Pattatucci, 1993a, 1993c). This does not mean that the results were wrong, but rather that they had a difference of opinion about how the population of gay men studied was identified and how the results were evaluated. This is real science. In some areas, and most notably for genetics, well-trained and competent scientists don't always agree on the right way to conduct a particular study or the right way to interpret the data. We will all have to wait to see which position will be validated by future studies.

A second linkage analysis from Hamer's group further examined male sexual orientation and markers within the Xq28 region of the X chromosome. Hu and colleagues evaluated a new set of 33 pairs of gay brothers and 36 pairs of lesbian sisters (Hu et al., 1995). In this study, gene sharing of markers within the Xq28 region of the X chromosome of homosexual-homosexual brothers or sisters was compared to gene sharing for homosexual-heterosexual brothers or sisters respectively. Gay brothers, that is homosexual-homosexual brothers, displayed sharing of markers at 67 percent, greater than the 50 percent level observable by chance, supporting their previous data demonstrating linkage of sexual orientation to the Xq28 region of the X chromosome. The rate of sharing observed for the gay-heterosexual brother pairs was 22 percent, and significantly less than the 50 percent sharing expected by chance. Taken together, these data replicate Hamer's earlier study. In addition, no such linkage was observed for lesbians, suggesting that a gene on the X chromosome may influence male, but not female sexual orientation. The data are shown in Figure 7.4. One thing that we must remember is that these results do not implicate an Xq28 linked gene in all male homosexuality, but only for 64 to 67 percent of those gay males whose **pedigree analysis** demonstrates maternal inheritance for the trait. Furthermore, the DNA of a fraction (17.5 percent in the first study and 22 percent in the second study) of those gay men whose family pedigree suggested maternal inheritance did not show linkage to the Xq28 region, suggesting that there may be other genes located on other chromosomes that influence male sexual orientation.

A second replication of Hamer's linkage study examined the sharing of markers at Xq28 in an independent sample of 54 pairs of gay brothers and found a 66 percent level of sharing (Sanders et al., 1998, cited in Hamer, 1999).

A third linkage analysis by Rice, Anderson, and colleagues examined linkage to Xq28 in male homosexuality, but found no evidence

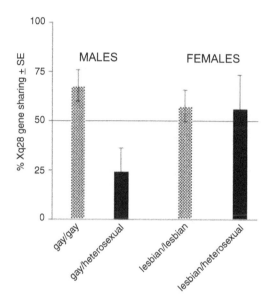

FIGURE 7.4. Hamer et al. data second study: Summary of gene sharing in male compared to female sib-pairs. The average levels of Xq28 gene sharing are shown for sib-pairs in which both members are homosexual and for sib-pairs in which one member is heterosexual and one member is homosexual and has the same Xq28 markers as his or her homosexual sibling. For males, n = 33 homo-sexual-homosexual sib-pairs, 12 homosexual-heterosexual sib-pairs; for fe-males, n = 36 homosexual-homosexual sib-pairs, 11 homosexual-heterosexual sib-pairs. *Source:* Hu et al. (1995).

to support the presence of a gene of large effect influencing male sex-ual orientation at position Xq28 (Rice, Anderson, Risch, & Ebers, 1999). While the authors claim that their data can be directly com-pared with that reported by Hamer's group, they admit that the popu-lation that they studied did not conform strictly to the selection crite-ria used by Hamer's group. Specifically, they did not select their families based on maternal transmission or inheritance. Their stated reason for not doing so was that they found no convincing evidence for such maternal transmission in the family pedigrees. While the Rice study did not support the presence of a gene at Xq28, they did state that their results do not rule out the possibility of detectable gene effects elsewhere in the genome on other chromosomes. So they agree with Hamer's group, that male sexual orientation is at least in part influenced by genes, but disagree about the presence of a gene or genes at Xq28 on the X chromosome.

A second technical comment by Hamer appeared in 1999 (Hamer, 1999) in which he presented a meta-analysis of all the available sexual orientation linkage data; two positive studies from his group, the positive study by Sanders et al. (1998), and the study by Rice, Anderson, et al. (1999) that failed to replicate the linkage of male sexual orientation to Xq28. The meta-analysis of all four studies showed an overall estimated sharing of 64 percent (Hamer, 1999). Further discussion of the differences between the Hamer and Rice et al. studies has been published (Rice, Risch, & Ebers, 1999).

So there you have it, a real scientific controversy. This difference of opinion is not just between two laboratories. In order to be published, each of the reports by Hamer's group and that by Rice's groups were reviewed by at least two, and probably more, competent working research geneticists. Furthermore, it is the practice of most researchers to ask their colleagues to review papers before they are submitted for publication. Therefore, it is probably safe to say that at least four to six or more scientists reviewed the papers with respect to the methods used, the data obtained, analysis, and the conclusions reached by the investigators. Each individual reviewer's questions, comments, and suggestions were no doubt carefully considered by the authors of each paper. So what we really have here is not a difference of opinion between two individuals or laboratories, but rather between two somewhat larger groups of professional geneticists. In this situation, the scientific community at large must simply wait for additional studies to confirm or refute the finding that male homosexuality is contributed to by a gene or genes on the X chromosome. So, for the moment, no one is right and no one is wrong; there is simply a difference of opinion, and only time and additional studies will determine if male homosexuality is associated with a gene or genes on the X chromosome.

Completion of the human genome project in 2003 provided geneticists with the DNA sequence of 20,000 to 25,000 genes. In addition, an array of new instrumentation and computer techniques were developed that allowed for a full genome scan of DNA from individuals displaying specific traits. The question that the application of these techniques can answer is, do individuals who display a particular trait have the same or nearly the same DNA sequences in specific areas of the genome relative to that observed for a large number of individuals who do not display the trait in question? Such a genome-wide scan of male sexual orientation has been conducted (Mustanski et al., 2005). A sample of 456 individuals from 146 families with two or more gay brothers was genotyped with 403 DNA markers. Three areas of significance were observed, one each on chromosomes 7, 8, and 10. These data can be used to focus future research on genes in these areas of chromosomes 7, 8, and 10. This study also found some additional evidence for linkage of some male homosexuality to Xq28.

Recent further work on the role of genes on the X chromosome influencing male sexual orientation has involved a comparison of extreme skewing of X chromosome inactivation in mothers of homosexual men. Each cell in a female embryo contains two X chromosomes. During development, each cell randomly inactivates one of the X chromosomes, and the inactive chromosome remains inactive in all daughter cells resulting from cell division. If we sample cells from a female, generally some cells will have one X chromosome inactivated, while others will have the other X chromosome inactivated. Through a series of procedures, one can determine the percentage of cells that have one of the two X chromosomes inactivated. Since inactivation is generally random, we might expect 50 percent of cells with one of the X chromosomes inactivated. There are circumstances that lead to nonrandom inactivation. Extreme skewing of X chromosome inactivation is said to occur when the number of cells with the same X chromosome inactivated exceeds 90 percent.

A recent study found that the number of women with extreme skewing of X chromosome inactivation was significantly higher in mothers of gay men than in age-matched control women without gay sons (Bocklandt, Horvath, Vilain, & Hamer, 2006). Skewing was greater for women with two or more gay sons. The statistical **significance** of the findings did not change if the control group was limited to women who had sons.

What does this mean? At present, we don't know, and the experiments need to be replicated. However, as suggested by the authors, these results support a role for the X chromosome in regulating some male sexual orientation. Furthermore, these results offer a path for further research on the **epigenetic** (i.e., factors that influence the expression of a trait, or phenotype, without a change in DNA sequence, or geneotype) basis of complex human traits.

SPECIFIC GENES

As we've seen (Chapter 4), both the androgen receptor and the enzyme aromatase are required for the sexual differentiation in mammals. One possibility is that DNA sequence variations in the androgen receptor gene could play a causal role in the development of male sexual orientation. DNA from a total of 197 homosexual males and 213 unselected subjects was examined using techniques designed to detect sequence variations in the androgen receptor. No significant differences in the sequence of the androgen receptor gene were found between homosexual men and the control group. Also, gay brothers did not share specific androgen receptor alleles (alternative versions of a gene that occupy a given site) (Macke et al., 1993).

The conversion of testosterone to estrogen by the enzyme aromatase in the brain is involved in the masculinization of the male behavior. In a separate study of 439 individuals from 144 unrelated families, no gene linkage was observed. Furthermore, no difference in the expression of aromatase messenger RNA was detected between nine homosexual men and eight heterosexual control subjects (DuPree, Mustanski, Bocklandt, Nievergelt, & Hamer, 2004).

As has been suggested, sexual orientation is a complex behavior that is not likely controlled by one gene. Furthermore, as the data presented demonstrate, multiple sites on autosomal chromosomes and the X chromosome show some relationship to some male sexual orientation. Thus, sexual orientation (both homosexual and heterosexual orientation) appears to be regulated by multiple genes. This is what would be expected for a complex behavioral trait. As suggested by Bocklandt et al. (2006), ". . . there might be several subgroups of gay men and women, each with their own specific biological origin" (p. 694).

SUMMARY

Experimental data support the conclusion that at least some male and female homosexuality is a consequence of genetics. This is supported by population studies and studies of identical twins. Homosexuality in both males and females runs in families. Furthermore, gay men have more gay brothers than do straight men, and lesbians have more lesbian sisters than do straight women. Recent data have identified areas of chromosomes 7, 8, and 10 as sites of particular interest with respect to a subpopulation of gay men (gay men with one or more gay brothers). No association between male homosexuality and the genes for either the androgen receptor or the enzyme aromatase was found. Several studies implicate the Xq28 area of the X chromosome in some male homosexuality. In addition, it appears that the X chromosome displays extreme skewing in the mother of some gay men. The genetics of female homosexuality has yet to be explored, but no association between Xq28 and lesbianism has been found.

Chapter 8

Brain Anatomy

In order to give you a complete picture of how much we know and do not know about sexual orientation and brain anatomy in humans, I will describe all of the articles published through 2006. As you will see, only a few studies are available. I present them all for completeness. For simplicity, the articles are presented in chronological order.

SEX-SPECIFIC BEHAVIOR PATTERNS

Before describing the neuroanatomical studies of human sexual orientation, it will be helpful to have some background information about neurons, the nervous system, and its complexity. The current view of brain function is called cellular connectionism. According to this view, individual neurons are the signaling units of the brain. Neurons are generally arranged in functional groups and connected to one another in a precise fashion. In order to understand the data to be presented, we must also know something about the general morphology or shape of neurons. Neurons are very specialized cells, but like all cells they contain a nucleus containing DNA and all of the other organelles required to support their function. Unlike other cells, however, neurons possess two cell-body extensions that other cells do not. The first of these extensions are the dendrites, which bring information from hundreds to thousands of other neurons into a single neuron. The second is the axon and its synaptic terminals, which is responsible for communicating with target cells.

Figure 8.1 shows the morphology of one type of neuron called a motor neuron. There are three general classifications of neuronal

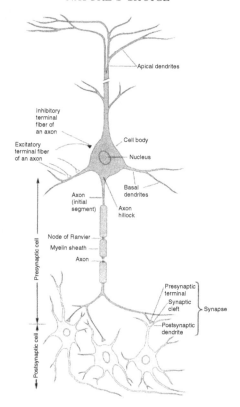

FIGURE 8.1. Morphology of neurons: Most neurons have several main features in common. The cell body contains the nucleus, the storehouse of chromosomes (DNA), the cell's genetic information. The cell body gives rise to two types of cell processes, the dendrites and the axon. Axons are the transmitting element of neurons and can vary greatly in length; some extend more than 1 meter (or 3 feet) within the body. Most axons in the central nervous system are very thin (between 0.2 and 20 μm in diameter, or much less than 1/1000 of an inch) as compared with the diameter of the cell body (50 μm or more, or about 2/1000 of an inch). Branches on the axon of one neuron (the presynaptic neuron) transmit signals to another neuron (the postsynaptic cell) at a site called the **synapse.** The branches of a single axon may form synapses with as many as 1,000 other neurons. The dendrites make up a major portion of the receptive surface of the neuron, and, together with the cell body, receive the synaptic input from presynaptic cells. *Source:* Adapted from Kandel, Schwartz, & Jessel (1995). Kandel, E.R., Schwartz, J.H., & Jessell, T.M. (Eds.), *Essentials of Neural Science and Behavior.* Copyright 1995. Appleton & Lange. Reproduced with permission of The McGraw-Hill Companies.

morphology, or cell shape. A neuron's morphology is generally unique, reflecting its location in the nervous system.

The neuron's job is to integrate all of the information received through its dendrites and cell body and respond with a specific signal transmitted through its axon to its target cell. In general, for a neuron that resides in the brain, all of its target cells are other neurons. The human brain is estimated to contain 100 billion neurons. The branches of a single axon may make contact with as many as 1,000 other neurons. So, the estimated number of neuronal connections in the brain is 100,000 billion, or 100 trillion. It is the most complex structure known.

In order to inquire about anatomical differences between brain areas of heterosexual versus homosexual individuals, we need to know where to look. We can get some clues about where to look by going back to the data obtained from animal studies. Once again, most of these studies were done on rodents. One easy first-order step is to simply ask which areas display differences between males and females. In doing this we identify what are termed *sexually dimorphic regions* of the brain. That is, these areas look different, or specific brain areas are of different sizes in male versus female animals. A difference in size suggests different numbers of neurons. An area larger in males than in females would be expected to contain more neurons. This still needs to be confirmed.

We know from studies of the brain areas in rodents that control ovulation and spermatogenesis that the hypothalamus is important for sexual function in both males and females. In fact, examination of the hypothalamus has revealed striking differences in the appearance of individual hypothalamic neurons in male versus female rodents. The nucleus of these neurons is larger in males than in females, indicating that the neuron is much more metabolically active in males. The density of the dendritic fields is greater for neurons in males than in females, indicating that more information is coming into the neuron from other neurons. Last, the size of the axons and the associated synaptic terminals is larger for neurons in males than in females, indicating a higher level of synaptic activity, or information going from the neuron to its target cells.

In addition, a specialized area of the hypothalamus called the sexually dimorphic nucleus of the preoptic area (SDN-POA) is greater in size in male than in female rodent brains (see Figure 8.2). That this

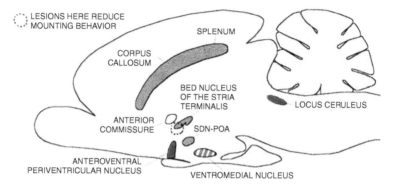

FIGURE 8.2. Rat brain sexually dimorphic nucleus: Sexually dimorphic nucleus of the preoptic area (SDN-POA) in the rat brain is among the regions whose size varies between males and females. Attempts to find an analogous cell group in humans have met with varying success. Some nuclei have not even been confirmed to exist in other rodents. Regions larger in males are shaded in light gray, and those larger in females are shaded in dark gray. *Source:* Byne (1994).

area is involved in sexual behaviors was demonstrated by transplanting the entire preoptic area from a newborn male into a female littermate's brain. As adults, these females displayed enhanced heterosexual and homosexual behaviors. The human hypothalamic structure analogous to the rodent SDN-POA is thought to be the interstitial nucleus of the anterior hypothalamus. We will examine the relationship of this structure with respect to human sexual orientation shortly.

These data indicate that adult sexual behaviors are at least in part controlled by neurons of the hypothalamus and that selected areas of the hypothalamus look different or are dimorphic between males and females. For this reason, a number of investigators interested in identifying sexually dimorphic areas with respect to orientation in humans have concentrated on areas of the hypothalamus. I will describe articles published through 2006 that deal with differences in brain structure size in homosexual versus heterosexual individuals.

SUPRACHIASMATIC NUCLEUS

The first report of a difference in brain structure related to sexual orientation in humans concerned a hypothalamic structure known as the **suprachiasmatic nucleus** (SCN) (Swaab & Hofman, 1990). This research article describes the measurement of the total volume of the SCN as well as the number of cells in the SCN of heterosexual and gay men. The SCN of gay men was found to be 1.7 times as large and contained 2.1 times as many cells as that of heterosexual men. Animal studies indicate that the SCN is the pacemaker that controls the body's natural day-and-night rhythms and is involved in **sexual reproduction.** It is however, not clear at present what role it plays in human sexual behavior, or how its function relates to sexual orientation. So, for now it is just an interesting observation.

THE ANTERIOR HYPOTHALAMUS

The next study addressed differences in hypothalamic structure and sexual orientation focusing on two areas of the anterior hypothalamus referred to as the interstitial nucleus of the anterior hypothalamus 2 and 3, or **INAH** 2 and INAH 3 (Le Vay, 1991). These areas, shown in Figures 8.3 and 8.4, were selected for study, because in nonhuman primates the anterior hypothalamus has been implicated in the generation of male-typical sexual behavior, and INAH 2 and 3 have been shown to be significantly larger in men than women (Allen, Hines, Shryne, & Gorski, 1989). Figure 8.4 gives you an appreciation of how close to one another all of the structures we are talking about in this chapter are. So, the question posed by the study was, do INAH 2 and/or INAH 3 differ in size with respect to sexual orientation?

To address this question, tissue containing INAH 1, 2, 3, and 4 were obtained from human brains at autopsy for three groups; 16 presumed heterosexual men, 19 gay men who died of complications of acquired immunodeficiency syndrome (AIDS), and 6 presumed heterosexual women. The heterosexual men and women were age-matched with the gay men. Data were obtained by staining very thin slices of tissue so that cells in each area could be visualized with a microscope (Figure 8.5). Each area was then traced into a computer. The computer added the area of the sections and determined the volume of each structure. The data represent the volume of each structure for each group. No significant differences in volume were found between the groups for INAH 1, 2, or 4. The average volume of INAH 3 for the three groups is shown in Table 8.1. The mean volume of INAH 3

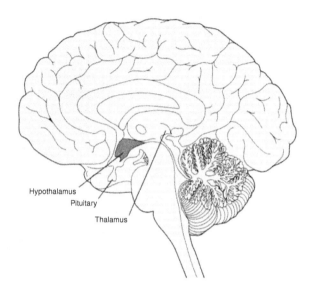

FIGURE 8.3. Hypothalamus of the human brain: Medial view of the brain showing the relationship of the hypothalamus to the pituitary and thalamus. *Source:* Kandel, Schwartz, & Jessel (1995). Kandel, E.R., Schwartz, J.H., & Jessell, T.M. (Eds.), *Essentials of Neural Science and Behavior.* Copyright 1995. Appleton & Lange. Reproduced with permission of The McGraw-Hill Companies.

found for heterosexual men is more than twice as large as that found for gay men and heterosexual women. Furthermore, the mean volume found for gay men was comparable to that found for heterosexual women. Thus, it is concluded that INAH 3 is dimorphic with respect to both sex and sexual orientation.

The possibility that the small size of INAH 3 in gay men was the result of AIDS or its complications and not related to the men's sexual orientation was not considered likely for the following reasons: (1) the size difference in INAH 3 was apparent even when comparing the gay men with heterosexual AIDS patients; (2) there was no effect of AIDS on the volume of the three other nuclei examined (INAH 1, 2, and 4); and (3) in the entire sample of AIDS patients there was no correlation between the volume of INAH 3 and the length of survival from the time of diagnosis.

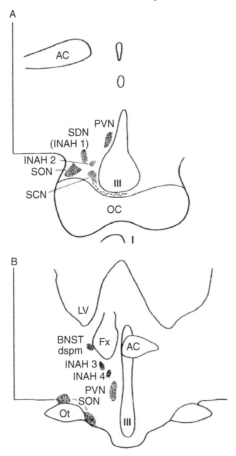

FIGURE 8.4. Topography of the sexually dimorphic structures in the human hypothalamus: A is a more rostral, or toward the front of the brain, view than B. Abbreviations: III, **third ventricle;** AC, anterior commissure; BNST-dspm, darkly staining posterior component of the bed nucleus of the stria terminalis; Fx, fornix; I, infundibulum; INAH1-4, interstitial nucleus of the anterior hypothalamus 1-4; LV, lateral ventricle; OC, optic chiasm; Ot, optic tract; PVN, paraventricular nucleus; SCN, suprachiasmatic nucleus; SDN, sexually dimorphic nucleus of the preoptic area; and SON, **supraoptic nucleus.** *Source:* Adapted from Swaab & Hofman (1995). Adapted from *Trends in Neurosciences,* Vol. 18, D.F. Swaab and M.A. Hofman, "Sexual Differentiation of the Human Hypothalamus in Relation to Gender and Sexual Orientation," pp. 264-270, copyright 1995, with permission from Elsevier.

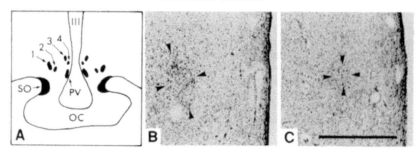

FIGURE 8.5. Histological sections of hypothalamic nuclei: (A) Diagram of a co-ronal section through the human hypothalamus at the level of the optic chiasm (OC). The four cell groups studied (INAH 1, 2, 3, and 4) are indicated by the cor-responding numerals. All four nuclei are not generally visible in the same coronal section. Supraoptic nucleus, SO; paraventricular nucleus, PV; and third ventri-cle, III. (B) Micrograph of INAH 3 from the left hypothalamus of a heterosexual male. The third ventricle is at the right of the figure. Arrowheads outline INAH 3. (C) Section from a homosexual male comparable to that in (B). INAH 3 is poorly recognizable as a distinct nucleus, but scattered cells similar to those constitut-ing the nucleus in heterosexual men were found within the area indicated by the arrowheads. The scale bar (1 mm, or 0.04 inches) applies to (B) and (C). *Source*: Le Vay (1991). Excerpted with permission from Le Vay, S. (1991). A dif-ference in hypothalamic structure between heterosexual and homosexual men. *Science, 253*(5023), 1034-1037. Copyright 1991 American Association for the Advancement of Science.

But, we must be careful how we use this data. The data are means, or averages. As can be seen in Figure 8.6, the range of the data is nearly two to four times the mean values. This is consistent with the general observation that for many attributes the differences among in-dividuals can be far greater than the average differences between the sexes. This means that if I told you that an individual had an INAH 3 volume of 0.073 mm³, you would not be able to look at the data and tell what sex they were or whether they were gay or straight. The data have no predictive value for individuals. All that can be said is that on average, a statistically significant difference exists in INAH 3 volume with respect to sex and sexual orientation.

As noted by Dr. Le Vay (1991, 1994), "the results do not allow one to decide if the size of INAH 3 in an individual is the cause or conse-quence of that individual's sexual orientation. . . ." (p. 1036). He also notes, however, that in rats the **sexual dimorphism** of the apparently

TABLE 8.1. Volume of INAH 3

Group	Volume	Number
Heterosexual men	0.12 ± 0.01 mm^3	16
Gay men	0.051 ± 0.01 mm^3	19
Heterosexual women	0.056 ± 0.02 mm^3	6

Source: Adapted from Le Vay (1991).

comparable hypothalamic structure, the sexually dimorphic nucleus of the preoptic area (SDN-POA), arises as a consequence of the dependence of its neurons on the circulating androgen (testosterone) levels during the critical period. Thus, although rats are not humans, it seems more likely that in humans the size of INAH 3 is also established early in life and later influences sexual behavior, rather than the other way around.

In 2001, Byne and collaborators published a second study of the size of the interstitial nuclei of the human anterior hypothalamus (Byne et al., 2001). With respect to INAH 3 and sexual orientation, they reported a nonsignificant trend toward INAH 3 occupying a smaller volume in gay men (14 subjects) than in heterosexual men (34 subjects). Furthermore, no difference in the number of neurons within INAH 3 was observed.

Most scientists agree that a single determination of a finding does not necessarily make it so. In order for a finding to be accepted as true, it must be replicated independently by another laboratory. Some of the determinations of dimorphism with respect to sex in areas of the hypothalamus have been replicated, and this suggests that the data concerning dimorphism with respect to sexual orientation may be valid. Table 8.2 shows all of the data available on sexual differences in the human hypothalamus (Mustanski, Chivers, et al., 2002). As you can see, some agreement and some disagreement exists. Many more studies will be required for us to achieve a clearer understanding of the sexual differences in human hypothalamic structures.

The hypothalamus is considered part of our primitive brain. Many lower animals have a hypothalamus, for example, reptiles. It is instructive to consider the behaviors regulated by the hypothalamus. The hypothalamus regulates sleep, thirst, eating, body temperature,

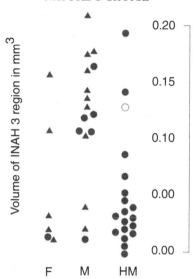

FIGURE 8.6. INAH 3 data: Volume of the hypothalamic nucleus INAH 3 for the three subject groups: females (F), presumed heterosexual males (M), and homosexual males (HM). Closed circles, individuals that died of complications of AIDS; closed triangles, individuals that died of causes other than AIDS; and open circle, an individual who was a bisexual male and died of AIDS. *Source*: Le Vay (1991).

emotional states, and reproduction, including sex. These are functions that we need to attend to in order to provide for normal physical and mental function. We do not have a choice about whether or not we need to attend to these functions. Our body sensations tell us we need to attend to them, and the only choice we have is how we will comply. We can delay compliance, but we cannot ignore it. Ignoring these functions leads to maladjustment and can lead to death. For example, we can delay eating, consuming liquids, and sleeping, but not indefinitely. Eventually we must satisfy these needs in order to survive. Failing to attend to our emotional states and our sexual needs also compromises our survival by leading to psychological maladjustment. Furthermore, one can make the argument that these functions are in place—"hardwired"—at birth, because they are essential for survival. This argument supports the conclusion that sexual orientation is fixed at birth and cannot be changed.

TABLE 8.2. Number of Studies with Positive and Null Findings in Neuroanatomical Research On Sex and Sexual Orientation

Brain Area	Sex Differences Positive Finding	Null Finding	Sexual Orientation Differences Positive Finding	Null Finding
SDN = INAH1	2: a,e	4: d,f,g,h	0	3: f,g,i
SCN	0	2: c,b	1: f	0
INAH2	1: d	2: g,h	0	2: g,i
INAH3	3: d,g,h	0	2: g,i	0
INAH4	0	3: d,g,h	0	2: g,i

Source: Mustanski, Chivers, & Bailey (2002). Reprinted with permission.
Note: Letters correspond to the following studies:
a = Swaab & Fliers (1985)
b = Swaab, Fliers, & Partiman (1985)
c = Hofman, Fliers, Goudsmit, Swaab, & Partiman (1988)
d = Allen, Hines, Shryne, & Gorski (1989)
e = Hofman & Swaab (1989)
f = Swaab & Hofman (1990)
g = Le Vay (1991)
h = Byne et al. (2000)
i = Byne et al. (2001)

An interesting statistic says that the average adult will experience 6,000 orgasms in his or her lifetime. Assuming that the average person becomes sexually active at age 15 and has a life expectancy of 75, this works out to about 100 orgasms per year, or about two orgasms a week, on average. The average woman is capable of reproduction from age 15 to about age 45, or for about 30 years. Based on the estimate of 6,000 orgasms in a lifetime, the average woman will have 3,000 sexual encounters capable of producing a pregnancy. In reality, the average woman could not manage more than one pregnancy a year, so she could theoretically have 30 children; however, most women would not survive 30 pregnancies. The point is, people engage in much more sex than can be justified on the basis of reproduction. Furthermore, half of the sex occurs after reproduction is possible (or for most of us, desirable). People don't stop having sex because they can no longer have children. So, the question is, what is all this sex for? Is sex just for fun? Do we do it because it feels good? Possibly. Two other mammals have sex for nonreproductive reasons, bonobos (pygmy chimpanzees) and dolphins (Diamond, 1997). Another way to approach this question is to ask when we are most likely to engage in sex, and what the major secondary benefit of sex is.

The major secondary benefit of sex is emotional stability. We are calmer after sex. We are more likely to have sex after something exciting happens, rather than after something sad or disappointing happens. If we finally graduate from college, get a raise, a new and better job, or win the lottery, we are excited. So in the midst of this bountiful excitement, nature has given us a way to calm down—sex. If conversely, we lose a job, get passed over for a promotion, or find bill collectors banging at our door, we are not excited but rather are closer to depression, and sex is seemingly the last thing on our minds.

If this is true, look how clever nature is. We engage in sex when we have a sense of competency and good fortune. These are conditions that would enable us to provide for a new offspring. When we are feeling down, overwhelmed, and not in control of our lives, we are not interested in sex, and we are not in a position to care for more than we have right now; no new babies please. Mother Nature has predisposed us to reproductive activity when we feel positive and excited, which also provides the secondary benefit of emotional stability.

Are there any other areas of the human brain that might be different between gay and straight individuals? The difficulty is knowing where to look. One study has looked at the **anterior commissure,** a structure outside of the hypothalamus, but closely adjacent to it.

THE ANTERIOR COMMISSURE

The next observation of dimorphism in brain structure with respect to sexual orientation was by Allen and Gorski (1992), and concerned the anterior commissure or AC. The AC is a **tract** of **axons** within the hypothalamus that primarily connects the right and left neocortex of the middle and inferior temporal lobes. In humans the AC allows for the transfer of visual, auditory, and olfactory information from one hemisphere to the other (Figure 8.7). Subjects consisted of 30 age-matched individuals for each group of heterosexual men, gay men, and heterosexual women. The data were obtained by measuring the cross sectional area of the AC at its midpoint and are shown in Table 8.3.

The area of the AC of gay men was significantly larger than that of heterosexual women (18.0 percent) and heterosexual men (34 percent). The area of the AC in heterosexual women was also significantly larger (13.4 percent) than that of heterosexual men. When the area of the AC in gay men without the two subjects with relatively large AC areas (see Figure 8.8) was compared with that of heterosexual men and heterosexual women, a significant difference remained in the area of the AC and in the area of the AC after adjusting for brain weight between heterosexual men, but not heterosexual women (Ta-

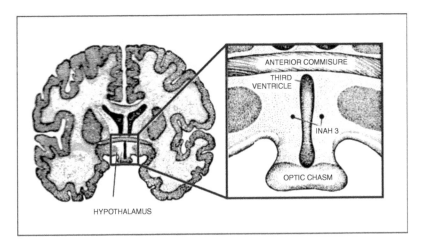

FIGURE 8.7. Section of human brain showing the anterior commissure: The location of the anterior commissure relative to the third ventricle, optic chasm, and INAH 3. *Source*: Le Vay & Hamer (1994).

TABLE 8.3. Midsagittal Cross-Sectional Area of the Anterior Commissure

Group	Area	Number
Heterosexual men	10.61 ± 0.5 mm^2	30
Gay men	14.20 ± 0.6 mm^2	30
Heterosexual women	12.03 ± 0.5 mm^2	30

Source: Adapted from Allen & Gorski (1992).

ble 8.4). Analysis of the data suggested no significant difference between men who did and did not die of AIDS.

From these data it can be concluded that the mean cross-sectional area of the AC at its midpoint is at least larger in gay men than in heterosexual men. However, if we look at the range of the values for each of the groups, we can see that knowing the cross-sectional area of the AC for an individual does not allow us to determine either their sex or sexual orientation (Figure 8.8). As with the volume of INAH 3 in the

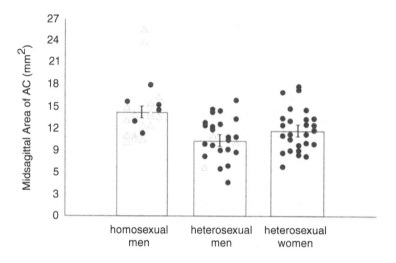

FIGURE 8.8. Computed midsagittal areas of the anterior commissure: Graph showing the computed midsagittal areas of the anterior commissure in 30 gay men, 30 heterosexual men, and 30 heterosexual women. Filled circles represent individuals without AIDS, and open triangles represent individuals who died of AIDS. *Source*: Allen & Gorski (1992).

TABLE 8.4. Midsagittal Cross-Sectional Area of the Anterior Commissure Corrected for Brain Weight

Group	Area/BW	Number
Heterosexual men	$7.4 \pm 0.4 \text{ mm}^2/\text{g} \times 10^3$	30
Gay men	$10.1 \pm 0.5 \text{ mm}^2/\text{g} \times 10^3$	30
Heterosexual women	$9.5 \pm 0.4 \text{ m m}^2/\text{g} \times 10^3$	30

Source: Adapted from Allen & Gorski (1992).

hypothalamus, the cross-sectional area of the AC is not predictive for either sex or sexual orientation.

At the time of publication, these data were considered important because they demonstrated a correlation between sexual orientation and the size of a brain structure not believed to be involved in sexual reproduction. As noted by Allen and Gorski (1992) at the time, this finding is important because it

> clearly argues against the notion that a single brain structure causes or results from a homosexual orientation. Rather, the correlation supports the hypothesis that . . . *factors operating early in development differentiate sexually dimorphic structures and functions of the brain in a* global *fashion*. (p. 7202)

In 2002, Lasco and collaborators published results of their study of the cross-sectional area of the AC in 121 subjects (20 gay men, 58 heterosexual men, and 43 heterosexual women) and found no variation in the area of the AC with respect to either sex or sexual orientation (Lasco, Jordan, Edgar, Petito, & Byne, 2002). Thus, at present whether the midsagittal area of the AC varies with respect to sexual orientation remains to be proven.

SUMMARY

We see that the brain is a very intricate organ, with 100 billion neurons and 100 trillion connections. However, through animal studies we know that the hypothalamus contains those structures containing the neurons and circuits involved in reproductive behaviors. We know that the preoptic area of the rodent hypothalamus is sexually dimorphic and larger in males than females. These studies have guided scientists in their exploration of comparable structures in humans. We do not have the technical means to examine the size of brain structures in live subjects. All of the studies reported here were done on autopsy samples of human brain.

The first study to report a difference in the size of a structure within the hypothalamus examined the suprachiasmatic nucleus (SCN). It is not clear how the SCN nucleus contributes to sexual orientation, but one study found the SCN of gay men to be 1.7 times larger, containing 2.1 times as many cells as that of heterosexual men.

The second study examined the interstitial nuclei of the anterior hypothalamus (INAH 1, 2, 3, and 4) in heterosexual men, gay men, and heterosexual women. These nuclei were of interest, because in nonhuman primates the anterior hypothalamus has been implicated in the generation of male-typical sexual behavior, and INAH 2 and 3 have been shown to be significantly larger in men than women. The study found that the volume of INAH 3 was more than twice as large in heterosexual men than in gay men and heterosexual women. Furthermore, the volume of INAH 3 in gay men was comparable to that of heterosexual women. It was concluded that INAH 3 was dimorphic with respect to sex and sexual orientation. An examination of all of the studies done on the hypothalamus reveals that there is some agreement and some disagreement. More studies will be required to clarify the relationships between the size of these structures and sexual orientation in humans.

We see that the hypothalamus regulates a number of behaviors including sleep, thirst, eating, body temperature, emotional states, and reproduction, including sex. The argument is made that these functions are so essential to survival that they are hardwired at birth. In addition, for human beings it is argued that sex serves more than just reproduction; it also seems to support emotional stability.

The last area examined with respect to sexual orientation is the anterior commissure (AC). The AC is closely adjacent to and just above the hypothalamus. The AC is a tract of axons that primarily connects the right and left neocortex of the middle and inferior temporal lobes. It allows for the transfer of visual, auditory, and olfactory information from one hemisphere to the other. In the study, the cross-sectional area of the AC at its midpoint was determined in heterosexual men, gay men, and heterosexual women. The area of the AC of gay men was significantly larger than that of heterosexual men and heterosexual women. The AC of heterosexual women was also significantly larger than that of heterosexual men.

The authors of the AC study noted that their finding

> clearly argues against the notion that a single brain structure causes or results from a homosexual orientation. Rather, the correlation supports the hypothesis that . . . *factors operating early in development differentiate sexually dimorphic structures and functions of the brain in a* global *fashion.* (p. 7202)

An additional study of the cross-sectional area of the AC in 121 subjects and found no variation in the area of the AC with respect to either sex or sexual orientation.

The positive findings conform to the neurohormonal theory in that these structures are dimorphic with respect to sexual orientation and gay men appear feminized with respect to the size of these structures. However, all three of these areas warrant further study before we can be confident that there are real size differences with respect to sexual orientation.

Chapter 9

Sex-Typical Behavior in Children

As we have discussed, sex-typical behaviors are those that are more prominent in one sex than the other. We see them in small children: rough-and-tumble play, competitive athletics, or aggression; toy and activity preference; imagined roles, careers, or role models; cross-dressing; preference for associating with male versus female individuals; and gender identity.

Several studies followed boys with marked patterns of childhood cross-gender sex-typical behavior from childhood to late adolescence when sexual orientation could be assessed (see Bailey & Zucker, 1995). In one study that examined fantasy or behavior in 66 feminine boys and 56 control boys, 75 to 80 percent of the previously feminine boys were either bisexual or homosexual at follow-up as assessed using the Kinsey rating scale as compared to 0 to 4 percent of the control boys. In a second series of studies (cite in Bailey & Zucker, 1995) 26 of 41, or 63 percent of, boys seen at follow-up had homosexual orientations. Thus, for males, data support a relation between patterns of childhood sex-typical behavior and later sexual orientation. This is true, particularly when individuals strongly favor opposite-sex behaviors—boys who exhibit strong feminine characteristics or girls who are tomboys.

A large meta-analysis of the relation between sex-typical behavior and sexual orientation evaluated data from 41 citations that assessed a total of 27,724 subjects; 8,963 heterosexual women, 1,729 lesbians, 11,298 heterosexual men, and 5,734 gay men (Bailey & Zucker, 1995). The meta-analysis revealed that for both men and women, homosexuals recall considerably more cross-gender sex-typical behavior in childhood than do heterosexuals. The result for homosexuals was more prominent for men than women. About 51 percent of boys who

Nature's Choice

display a degree of cross-gender sex-typical behavior that is typical for those who will become homosexual actually do become homosexual. For women, the corresponding percentage is 6 percent. Thus, childhood cross-gender sex-typical behavior is predictive of adult homosexuality.

According to the neurohormonal theory, sexual orientation is associated with the establishment of permanent differences in the limbic areas of the brain, particularly the hypothalamus. This occurs from the middle of the second to the end of the fifth month of gestation. Sex-typical behaviors are associated with diverse areas of the brain, extensively involving the cortex, and are established during a period that overlaps the period for the establishment of sexual orientation and extends beyond it for two to three more months. Thus, the critical periods for sexual orientation and sex-typical behavior determination are distinct, but partially overlapping. Therefore, fluctuation in androgens and other relevant hormones would likely influence both sexual orientation and sex-typical behaviors, but could also affect only one, if androgen availability were limited with respect to gestational time.

Support for this idea has come from a study of exposure of female monkeys (rhesus macaques) to testosterone during gestation (Goy, Bercovitch, & McBrair, 1988). Female monkeys exposed to testosterone early in gestation showed increased rates of maternal and peer-mounting (male-typical behaviors), but did not differ in their rates of rough-and-tumble play relative to normal control females. In contrast, females exposed to testosterone late in gestation showed increased rates of rough-and-tumble play and peer mounting, but did not differ in their rates of maternal mounting relative to normal control females. The authors concluded that ". . . the individual behavior traits that are components of the juvenile male role are independently regulated by the organizing actions of androgen and have separable critical periods" (p. 552).

Some insight into the effects of exposure of human females to androgens during gestation comes from studies of girls with congenital adrenal hyperplasia (CAH). Girls with CAH are exposed to elevated levels of androgens during gestation and show more masculine-typical behaviors as children; altered behaviors include an increased preference for boys' toys, boys' activities, and boys as playmates. Past attempts to determine whether the normal variability in hormone exposure during gestation contributes to human behavioral sex differences have produced inconsistent results.

A recent study of a large number of mothers and their children has asked if a correlation exists between the levels of testosterone in mother's blood and the male-typical behavior of their children, both girls and boys (Hines et al., 2002). Participants were part of the Avon Longitudinal Study of Parents and Children in the U.K. and consisted of 13,998 women who gave birth to 14,138 children between April 1991 and December 1992. Data from 679 children, 337 girls, and 342 boys were analyzed for the study.

Sex-typical behavior of each child was assessed at age 3.5 years using the Pre-School Activities Inventory (PSAI), a questionnaire measure on which parents indicate their child's involvement in various sex-typed behaviors (Golombok & Rust, 1993a,b). The PSAI is a standardized and reliable screening instrument and is specifically designed to differentiate "masculine" and "feminine" boys and girls within a normal population sample of preschool children. It differentiates within, as well as between, the sexes.

Levels of testosterone were determined in maternal blood samples obtained at 8 to 24 weeks of pregnancy. The maternal levels of the testosterone-binding globulin, sex-hormone-binding globulin (SHBG), which limits the ability of testosterone to act, were also determined.

Data were analyzed using correlation coefficients and analyses of covariance. Background variables included in the analyses of covariance were (1) maternal education, (2) the presence of older brothers or sisters in the home, (3) the presence of a male partner living with the mother in the home, and (4) parental adherence to traditional sex roles.

Data indicated that the levels of testosterone and SHBG were not different for pregnancies producing either male or female offspring. For girls, the data indicated a significant and positive, linear relationship between maternal testosterone and masculine-typical gender role behavior; as the level of testosterone increased so did masculine-typical behavior. This would seem to suggest a dose-dependent relationship between testosterone levels and sex-typical behaviors. Maternal testosterone during pregnancy accounted for approximately 2 percent of the variance in the gender role behavior of preschool girls. There was no effect of SHBG on behavior in girls. No relationship between testosterone or SHBG and sex-typical behavior was observed for boys.

The findings for girls were further examined with respect to background variables. None of the background variables correlated with either testosterone or male-typical behavior and none altered the correlation between testosterone and male-typical behavior in girls. Thus, the background variables did not account for the observed relation between testosterone and sex-typical behavior as indicated by PSAI scores in preschool girls. It should be noted that total serum testosterone in prepubertal boys and girls are nearly the same (8 to 14 nanograms per milliliter for boys and 5 to 13 nanograms per milliliter for girls).

This study reports a correlation between the maternal blood testosterone levels during pregnancy and the sex-typical behavior of preschool female offspring. The data support a correlation between the variables and not a cause-and-effect relationship. This means that other biological and/or environmental (that is social and cultural) variables not under control in the experiment could be causative. Nonetheless, these data answer the posed question of whether a correlation exists between maternal testosterone levels and childhood male-typical behaviors in both boys and girls. The answer is yes for girls and no for boys.

SUMMARY

The basic question examined in these studies is whether childhood sex-typical behavior is related to the expression of sexual orientation in adulthood. Studies followed feminine and control boys and found that boyhood femininity predicted a homosexual orientation. Furthermore, a large meta-analysis assessed recalled childhood behavior and found that for both men and women, homosexuals recall more cross-gender sex-typical behavior in childhood than do heterosexual. We also see that a study in primates suggests that ". . . the individual behavior traits that are components of the juvenile male role are independently regulated by the organizing actions of androgen and have separable critical periods" (Goy, Bercovitch, & McBrair, 1988, p. 552).

Girls with congenital adrenal hyperplasia (CAH) are exposed to elevated androgens during gestation and show more masculine-typical behaviors as children. Also, women with CAH recall more male-typical play behavior as children than do unaffected women.

A large study also examined the relationship between maternal circulating levels of testosterone and the play behavior of their male and female offspring at 3.5 years of age. For girls, the data revealed that as the level of testosterone increased in maternal circulation, the male-typical play behavior of their daughters did also. No relationship between maternal testosterone and sex-typical behavior was observed for boys. Once again, these finding are consistent with predictions of neurohormonal theory.

Chapter 10

Anthropometrics: Body Measurements

Anthropometrics is the comparative study of human body measurements. The general approach for these studies is to assess differences between heterosexual and homosexual populations on those measurements that are sexually dimorphic, that is, different between men and women. Sexual dimorphism in skeletal size and shape emerges during childhood and accelerates during adolescence (see Martin & Nguyen, 2004 and references cited therein). A number of body measurements have been evaluated with respect to sexual orientation (see Mustanski, Chivers, et al., 2002 for a review), including long bones, finger-length ratio, dermatoglyphics, and specifically the analysis of the number of skin ridges of fingerprints, height, weight, waist-to-hip ratio in women, and penis size. In addition, the onset of puberty, which is sexually dimorphic, with females reaching puberty earlier than males, has been studied with respect to sexual orientation.

LONG BONES

When long bone development is complete, dimorphism is marked in the long bones of the arms and legs relative to the arm span or stature; males have greater arm:span and leg:stature ratios than do females. Postnatal development of this skeletal dimorphism is under the control of **gonadal steroids,** primarily testosterone and estrogen, and growth hormone. Steroid hormones influence bone growth directly and by stimulating growth hormone secretion. The relative size of bones in adulthood reflects the pattern and levels of steroid hormone and growth hormone action during childhood and may also re-

Nature's Choice

flect prenatal steroid action. Therefore, as reasoned by Martin & Nguyen (2004), anthropometric analysis of adult males and females may be useful in identifying variations in the level of androgen and estrogen receptor activation during development. To this end, these investigators made extensive bone measurements on individuals of Caucasian ancestry between the ages of 20 and 50 (118 heterosexual men, 116 gay men, 60 lesbians, and 109 heterosexual women). Height for women and men did not vary with sexual orientation, although there was a trend for lesbians to be taller than heterosexual women and gay men to be shorter than heterosexual men. Weight and BMI of heterosexual men was greater than that of gay men. Similarly, the weight and BMI of lesbians was greater than that for heterosexual women. Similar results have been reported. Hand measurements revealed that the width:length ratio was greater for heterosexual men relative to gay men and greater for lesbians relative to heterosexual women. Similarly, hand length was greater for heterosexual men than for gay men by 1.41 mm, and greater in lesbians than heterosexual women by 2.2 mm. Heterosexual men and lesbians had significantly longer legs and greater long bone growth in the arms than did gay men and heterosexual women respectively. It was found that overall, both gay men and lesbians were approximately 33 percent closer to the opposite sex in arm:stature (arm:height) ratios. Overall, these data conform to the notion that gay men are feminized relative to heterosexual men and lesbians are masculinized relative to heterosexual women. The authors concluded that their data suggest either different levels of hormonal signals in homosexual and heterosexual during childhood, or different sensitivities of target tissues induced by an earlier perinatal hormone exposure, or both.

FINGER-LENGTH RATIO

The development of the urinogenital system, which includes the gonads and genitalia, and the appendicular skeleton (hands and feet), are under the control of the same genes, *Hoxa* and *Hoxd,* which are members of the class of genes called Hox genes (Kondo, Zakany, Innis, & Duboule, 1997). In mice, deregulation of *Hoxd* gene expression may alter the relative lengths of digits and affect the growth of the genital bud (Kondo et al., 1997). In humans, anatomical defects in digits and genitalia occur in the hand-foot-genital syndrome, which results from mutations in *Hoxa* (Mortlock & Innis, 1997). The common control of digit and gonad differentiation suggests that patterns of digit formation may relate to hormonal concentrations, and that this control is likely exerted during gestation. Sex differences have been reported for the relative lengths of toes in mice (Brown, Finn, & Breedlove, 2001; Manning, Callow, & Bundred, 2003) and for both the metacarpals (the bones in the hand extending from the wrist to the fingers) and metatarsals (the bones in the foot extending from the ankle to the toes) of baboons (McFadden & Bracht, 2002a), gorillas, and chimpanzees (McFadden & Bracht, 2002b). Finger-length measurements are taken from ink or photocopy palm-down handprints. The ratio of the length of the second to fourth finger, or finger-length ratio (2D:4D ratio), is sexually dimorphic in humans; in women, the second digit or index finger (2D) is about equal in length to the fourth digit or ring finger (4D), with the 2D:4D ratio equal to 1.00, but in men, 2D is shorter on average than 4D, with the 2D:4D ratio equal to 0.98 (George, 1930; Phelps, 1952; Manning, 2002). Studies of fetal material suggest that the adult relative finger length is established in utero by week 13 (about 3.5 months) (Garn, Burdi, Babler, & Stinson, 1975) and reaches its term length in the last trimester (Malas, Dogan, Evcil, & Desdicioglu, 2006). Furthermore, cross-section evidence shows that the 2D:4D ratio is fixed at least as early as 2 years and shows no significant change at puberty or thereafter (Phelps, 1952; Manning, Scott, Wilson, & Lewis-Jones, 1998).

A 2D:4D ratio, less than 1, is typical of men and associated with high levels of androgens, while in contrast, a 2D:4D ratio greater than 1 is associated with women and high estrogen levels (Manning et al., 1998). Additional studies support the conclusion that a low 2D:4D ratio is related to high testosterone levels. Observations include: (1) a low 2D:4D ratio correlates with high **sperm** count, high testosterone levels, enhanced male cognitive skills, and athletic ability, and (2) a high 2D:4D ratio correlates with low testosterone levels and high estrogen levels as indicated by high female fertility markers, including family size, waist-to-hip ratio, enhanced female cognitive skills, and diseases such as breast cancer (Lutchmaya, Baron-Cohen, Raggatt, Knickmeyer, & Manning, 2004; Manning, 2002; Manning & Leinster, 2001; Manning et al., 1998; Manning & Bundred, 2000; Manning & Taylor, 2001; Manning, Trivers, Singh, & Thornhill, 1999). The most convincing data relating low 2D:4D ratios to prenatal androgen levels comes from a study of males and females with CAH (congenital adrenal hyperplasia), who as a consequence are exposed to high levels of androgens in utero (Brown, Hines, Fane, & Breedlove, 2002). Brown, Hines, et al. (2002) found that females with CAH had significantly smaller, that is more masculine-typical 2D:4D ratios on the right hand than did females without CAH. No difference was found between females with and without CAH in the absolute length of any of the four fingers measured, only in the ratio of the length of 2D to 4D. Males with CAH had significantly smaller 2D:4D ratios on the left hand than did males without CAH. Furthermore, a subset of males with CAH had significantly smaller 2D:4D ratios on both hands compared to their male relatives without CAH. These results are consistent with the idea that prenatal exposure to androgens reduces the 2D:4D ratio and establishes the sex difference in human finger-length patterns. The relationship between 2D:4D ratio and prenatal testosterone is further supported by the recent observation that the 2D:4D ratio is masculinized in the female of opposite sex twins compared to that for female twins (van Anders, Vernon, & Wilbur, 2006).

Additional evidence for the probable causal relationship between prenatal androgen levels and the sexually dimorphic nature of the 2D:4D ratio comes from two additional studies. First, low 2D:4D ratios are associated with high amniotic concentrations of testosterone relative to estrogen (Lutchmaya et al., 2004). Second, variations in the androgen receptor gene sequence that increase the sensitivity to testosterone are related to low, that is, masculinized, 2D:4D ratios (Manning, Bundred, Newton, & Flanagan, 2003).

The general methods employed by these studies are very similar. Participants were recruited through publications, college student classes, and organizations. Their sexual orientation was assessed either through an investigator-devised questionnaire or by use of the Kinsey Scale (Kinsey et al., 1948) and/or the Klein Grid (Klein, Sepekoff, & Wolf, 1985). In some cases extensive family and demographic data were obtained from each subject. Palm down handprints were obtained as ink-prints or photocopies. Finger-length measurements were usually made using a caliper. Finger-length measurements were often repeated by a second investigator in order to determine the reliability of the measurements. The 2D:4D ratio was then determined by dividing the length of the second digit by the length of the fourth digit.

The 2D:4D finger-length ratio of men and women with respect to sexual orientation has been examined in no fewer than a dozen studies since 1998. As shown in Table 10.1, most of these studies confirmed that the 2D:4D ratio is sexually dimorphic, that is, smaller in men than women. The results of the studies of the 2D:4D ratio in homosexual versus heterosexual men and women are summarized in Table 10.1.

2D:4D Ratio in Women

Data obtained thus far on the relationship between the 2D:4D ratio and sexual orientation in women appear straightforward. Examining Table 10.1 we see that six out of eight studies found the 2D:4D ratio for lesbians to be statistically smaller than that for heterosexual women and comparable to that for heterosexual men. Doing a little limited arithmetic, the mean value for the 2D:4D ratio for lesbians across these studies is 0.965. This is very close to the mean value for heterosexual men derived from these studies, or 0.958. Another interesting comparison is that between the values for lesbians and that for females with CAH, which is 0.956 (Brown, Finn, et al., 2002; Brown,

TABLE 10.1. Studies of 2D:4D Finger Length Ratio with Positive or Null Findings

Groups	Positive	Null
Men < Women	8[a]	1[b]
Lesbians < HetW	6[c]	2[d]
GayM < HetM	2[e]	
GayM = HetM		4[f]
GayM > Het M	3[g]	

[a]Manning, Scott, Wilson, & Lewis-Jones (1998); Williams et al. (2000); McFadden & Shubel (2002); Brown, Finn, et al. (2002); Brown, Hines, et al. (2002); Lippa (2003a); Kraemer et al. (2006); Manning, Churchill, & Peters (2007).
[b]Rahman & Wilson (2003b).
[c]Williams et al. (2000); McFadden & Shubel (2002); Brown, Finn, et al. (2002); Rahman & Wilson (2003b); Hall & Love (2003); Kraemer et al. (2006).
[d]Lippa (2003a); Manning, Churchill, & Peters (2007).
[e]Robinson & Manning (2000); Rahman & Wilson (2003b).
[f]Williams et al. (2000); Brown, Hines, et al. (2002); Voracek, et al. (2005); Kraemer et al. (2006).
[g]McFadden & Shubel (2002); Lippa (2003a); Manning, Churchill, & Peters (2007).

Hines, et al., 2002). Once again, the values are very close. These data support the conclusion that lesbians may well have been exposed to elevated levels of androgens during gestation.

One study (Lippa, 2003a) did not find the 2D:4D ratio for lesbians to be smaller than that for heterosexual women, but rather the ratio for lesbians was larger than that for heterosexual women. It is interesting that the reported ratio for lesbians of 0.960 is very close to the rough mean calculated previously. Thus, the conclusion that the ratio for lesbians is greater than that for heterosexual women is based on the ratio determined for heterosexual women (0.953), which is much lower than that reported by others. Throughout these studies it has become apparent that the 2D:4D ratio is influenced by race and ethnicity. Some of these studies did not take this into account while others did. In general it was not a large problem since the groups studied were predominantly Caucasian. The most recent study by Manning, Churchill, & Peters (2007) did look at the 2D:4D ratio across five different ethnic groups. Although a trend was found for lesbians to have a lower 2D:4D ratio than heterosexual women across all five ethnic groups (except for the right hand ratios for Chinese women), the differences were not statistically significant.

The study by Hall & Love (2003) looked more directly at the influence of genes versus prenatal environmental factors in the determination of 2D:4D ratios between lesbians and heterosexual women. To do this, they studied seven sets of monozygotic (identical) female twins discordant for sexual orientation (one was homosexual and one was heterosexual), and five sets of female identical twins that were concordant for sexual orientation (both twins were homosexual). Twins were compared only to their co-twin. In the twins of opposite sexual orientation (seven sets), statistically significant differences were found for the 2D:4D ratio of both hands. In the five sets of twins in which both were homosexual, no statistically significant differences in the 2D:4D ratio were found for either hand. The average ratios for the lesbians in this study was 0.98 to 0.99 and was consistent with the population average for European males, while the average for the heterosexual women was 1.00 to 1.01, consistent with values determined for European females (Manning et al., 1998). The authors concluded that this study supports previous findings that lesbian sexual orientation is associated with lower 2D:4D ratio, and, by extrapolation, higher androgen levels during prenatal development.

They acknowledge that these results need to be viewed with caution because of the small sample size, but point out that the study of monozygotic twins allows for differences in 2D:4D ratio to be interpreted primarily as due to prenatal environmental factors. The differences observed between co-twins indicate that differences existed in the prenatal environment during the first trimester of development. Furthermore, these differences are associated with sexual orientation, suggesting that prenatal environmental factors should be considered as causative factors in its development.

2D:4D Ratio in Men

Examining Table 10.1 again we see that the story for men is more complicated at present, as there is no clear consensus. We have two studies that conclude that the 2D:4D ratio for gay men is smaller than that for heterosexual men, four studies that found no difference in the 2D:4D ratio between gay men and heterosexual men, and three studies that found that the 2D:4D ratio was greater for gay men than for heterosexual men. The explanation for the result that the 2D:4D ratio for gay men is smaller than that for heterosexual men is that gay men are hypermasculinized relative to heterosexual men with respect to this trait. It is further suggested that this indicates that they were exposed to more testosterone than heterosexual men during gestation. Conversely, where the 2D:4D ratio for gay men is larger than that for heterosexual men, and thus intermediate between the ratio for heterosexual men and heterosexual women, the explanation is that the gay men are hypomasculinized relative to heterosexual men. The interpretation here is that the gay men were exposed to lower levels of testosterone than heterosexual men during gestation. The picture for men is even more complex, as there appears to be significant effects of ethnicity on the 2D:4D ratio. This fact first emerged in a reanalysis of the data from five previous studies (McFadden et al., 2005). The reanalysis revealed that the constancy of the 2D:4D ratio for white homosexuals did not extend to homosexuals of three other ethnicities. In a larger study of this issue involving a vary large number of subjects (104,688 individuals), higher 2D:4D ratios were found in gay men and bisexual men relative to heterosexual men of Caucasian background (particularly white), but not for men of black or Chinese backgrounds (Manning et al., 2007). At present there is no understanding of the effects of ethnicity on the differences in 2D:4D ratios between gay men and heterosexual men. The large number of subjects in this study provides additional support for gay men having larger 2D:4D ratios, that is, they are hypomasculinized relative to heterosexual men. What has not as yet been examined is the 2D:4D ratio in gay men who owe their sexual orientation to the fraternal birth order effect. Nor do we know if testosterone plays a role in the fraternal birth order effect.

In summary then, we have six studies that concluded that the 2D:4D finger-length ratio for lesbians is lower than that for heterosexual women and two studies that concluded that there is no difference in the 2D:4D ratio between lesbians and heterosexual women. One of the positive studies provided data on monozygotic twins, where intra-twin comparisons revealed that the lesbian twin's 2D:4D ratio was lower than her heterosexual sister's ratio. Furthermore, the provisional mean value for the 2D:4D ratio of lesbians is comparable to that reported for females with CAH, providing support for a relationship between female homosexual orientation and exposure to elevated levels of androgens during gestation. These findings support the conclusion that exposure to androgens during gestation influences the 2D:4D ratio and sexual orientation of at least those lesbians who self identify as expressing male-typical behaviors (i.e., they describe themselves as "butch").

At present no conclusion with regard to the 2D:4D ratio in gay men versus heterosexual men appears justified although the recent large studies appear to support a 2D:4D ratio for gay men greater (hypomasculinized, or feminized) than that for heterosexual men.

DERMATOGLYPHICS: FINGERPRINT ANALYSIS

Dermatoglyphics is the study of the patterns of the skin ridges on the fingertips (fingerprints), palms, toes, and soles of all primates. In humans, the number of ridges and their pattern are determined between the 8th and 16th week of fetal life (Holt, 1968). After birth, ridge counts and patterns are stable and not affected by development or the environment; they can be altered only by severe mechanical damage. Thus, because they develop during gestation and are stable after birth, those traits that show a correlation with them are likely to have the same developmental timing, occurring during mid-gestation.

The dermal ridges of fingerprints are strongly influenced by genetics, but are subject to local, intrauterine environmental alterations during the critical period of their development. Dermal ridge counts can however be altered by maternal use or treatment with drugs. The total ridge count is a highly heritable human trait, with a correlation of 0.95 or greater between monozygotic (identical) twins (Bouchard, Lykken, McGue, Segal, & Tellegen, 1990). In nonhuman primates, psychological stress has been shown to affect ridge patterns (Newell-Morris, Fahrenbruch, & Sackett, 1989). In addition, in nonhuman primates, increased levels of testosterone administered during gestation lowered total dermal ridge counts relative to controls (Jamison, Jamison, and Meier, 1994). In humans, a correlation has been shown between adult testosterone levels and greater leftward dermatoglyphic asymmetry (Sorenson-Jamison, Meier, & Campbell, 1993). Leftward asymmetry is defined as two or more ridges on the left hand; conversely, rightward asymmetry would reflect two or more ridges on the right hand. Whether this reflects a difference in the levels of testosterone during gestation remains to be determined.

Last, the total number of ridges on both hands is sexually dimorphic, with men tending to have a higher total ridge count than women (Dittmar, 1998; Hall & Kimura, 1994; Micle & Kobyliansky, 1988). This difference in total ridge count is not due to the large hand size of men (Holt, 1968).

Four studies have examined ridge asymmetry and/or total ridge counts in homosexuals. The general procedures used by these studies included sexual orientation assessment by self-declaration and completion of the Kinsey Scales and/or the Klein Grid and inked fingerprints to assess ridge asymmetry and total ridge counts.

The first study to appear by Hall & Kimura (1994) determined directional asymmetry and total ridge counts in 182 heterosexual men and 66 gay men. According to the authors, ridge patterns on the 3 middle fingers have no tri-radial point where the ridge count is 0. For this reason, only the thumb and little finger of both hands were scored for total ridge counts. No significant difference in total ridge count was detected between heterosexual and gay men. On the other hand, a statistical comparison of sexual orientation and the direction of ridge asymmetry indicated that more gay men possessed the minority leftward asymmetry than did heterosexual men.

The second study by Hall (2000a, 2000b) assessed total ridge count in female monozygotic twins discordant for sexual orientation (one twin was heterosexual and the other was lesbian). As the author states, since dermatoglyphic characteristics do not change after the second trimester, differences in dermatoglyphics between genetically identical twins can be used as a marker for prenatal environmental differences. In contrast to the study by Hall & Kimura (1994), total ridge counts were determined for all fingers on both hands. In a set of 7 female twins discordant for sexual orientation, all of the lesbian twins had statistically significant lower total ridge counts than their heterosexual twin. In a control group of 5 sets of twins concordant for sexual orientation (both twins were homosexual) total ridge counts were not significantly different between twins.

The third study by Mustanski, Bailey, & Kasper (2002) examined a number of factors including total ridge count and directional asymmetry. This was the largest of the three studies, examining 169 gay men, 164 heterosexual men, 117 lesbians, and 164 heterosexual women. As in the study by Hall & Kimura (1994), these investigators made total ridge counts on only the thumb and little finger, because, as stated by the authors, the three middle fingers have a higher incidence of arches; this results in a score of zero. No significant difference was found for any of the dermatoglyphic features between the homosexual and heterosexual groups studied.

The fourth study (Forastieri et al., 2002) determined directional asymmetry and total ridge count for 60 gay men, 76 heterosexual men, and 60 heterosexual women. A prevalence of rightward asymmetry was observed for all 3 populations. No significant difference in leftward asymmetry or total ridge count was observed between gay men and heterosexual men.

A difference exists between men and women in total ridge count, with men having a higher total ridge count. Furthermore, data suggest that testosterone can influence total ridge count. Table 10.2 summarizes the dermatoglyphic results reported to date. Only four studies have been published, but there appears to be consensus among them that no difference in total ridge count exists between gay men and heterosexual men. Two studies also agree that no difference in the directional asymmetry of fingerprints exists between gay men and heterosexual men.

TABLE 10.2. Summary of Dermatoglyphic Studies

Groups	Pattern	Total Counts
GayM vs. HetM[a]	+, left*	No difference
GatM vs. HetM[c]	No difference	No difference
GayM vs. HetM[d]	No difference**	No difference
Lesbians vs. HetW[c]	No difference	No difference
Female Twins[b]		Lesbians < HetW

*Gay men displayed greater leftward asymmetry.
**Rightward asymmetry in all groups, including heterosexual women.
[a]Hall and Kimura (1994).
[b]Hall (2000a,b).
[c]Mustanski, Chivers, & Barley (2002).
[d]Forastieri et al. (2002).

Only two studies (Hall, 2000b; Mustanski et al., 2002) of women have been reported, each using a different experimental design and a different assessment procedure; however, the results of the two studies do not agree. Studies of monozygotic twins, who have identical genes, have considerable power because they can assess the influence of factors other than genetics. In addition, the study (Hall, 2000b) made comparisons between only twins of the same sex, and not between groups of genetically unrelated individuals. The reported study (Hall, 2000b), however, was of a very small number of twins, and, as the author states, due to the small sample size, the results should be interpreted carefully. The study (Hall, 2000b) reported lower total ridge counts for lesbians as compared to their heterosexual twins. In a separate set of twins, each of the same sexual orientation, no difference in total ridge count was observed. A second study (Mustanski et al., 2002) that included women did not find that lesbians had lower total ridge counts than heterosexual women. So, at present, there is no consensus on the dermatoglyphic traits for women with respect to sexual orientation.

HEIGHT AND WEIGHT

The rationale for examining height and weight with respect to sexual orientation comes from two factors. First, these characteristics are sexually dimorphic, with men being on average taller and heavier than women. Second, these traits could be influenced by sex-atypical organization of sexually dimorphic brain structures, including the hypothalamic-pituitary-gonadal axis, that control physical growth (Blanchard & Bogaert, 1996a).

The neurohormonal theory would predict that gay men would be feminized with respect to height and weight and would therefore be shorter and lighter than heterosexual men. Conversely, lesbians would be expected to be masculinized on these measures and would therefore be taller and heavier than heterosexual women.

Two studies involving a total of 484 gay men and 564 heterosexual men found gay men to be shorter than heterosexual men (see Table 10.3). The most rigorous study of height in men to date found gay men to be 1.5 cm (0.59 inches) shorter than heterosexual men (Bogaert & Blanchard, 1996). An additional three studies of 412 gay men and 7,983 heterosexual men found no difference in height between the two groups.

Only two studies of height in women have been conducted (see Table 10.3). The first study of 148 lesbians found a statistically non-significant difference of 0.4 inches in height between the lesbians and the published results (Heath, Hipkins, & Miller, 1961) of 2,434 heterosexual women (Perkins, 1981). When the lesbian population was classified according to psychosexual behavioral categories of dominant, intermediate, and passive, differences in height were significant. The dominant group was significantly taller than the passive group. Furthermore, on several measures of physique, the passive lesbian group most closely resembled heterosexual women. Thus, it seems likely that further studies of lesbians classified according to

TABLE 10.3. Summary of Height Studies

Finding	Number of Studies
GayM shorter than HetM[a]	2
GayM same height as HetM[b]	3
Lesbians taller than HetW[c]	1
Lesbians the same height as HetW[d]	1

[a]Blanchard, Dickey, & Jones (1995); Bogaert and Blanchard (1996).
[b]Evens (1972); Blanchard and Bogaert (1996a); Bogaert and Friesen (2002).
[c]Bogaert (1998a).
[d]Perkins (1981).

psychosexual categories will demonstrate a significant difference in height between dominant lesbians and heterosexual women. The more recent and rigorous study of 275 lesbians and 5,201 heterosexual women found lesbians to be 0.89 cm (0.35 in) taller than heterosexual women (Bogaert, 1998b).

Weight differences between gay men and heterosexual men have been examined in a total of 7 studies (see Table 10.4). Four studies that examined a total of 1,382 gay men and 4,779 heterosexual men found that gay men weighed less than heterosexual men. In contrast, 3 additional studies involving 157 gay men and 5,263 heterosexual men found no significant difference in weight between the two groups.

Weight differences between lesbians and heterosexual women have been examined in five studies (see Table 10.4). Four studies involving a total of 563 lesbians and 10,723 heterosexual women found lesbians to be heavier than heterosexual women. One study found no difference in weight between lesbians and heterosexual women. A recent study of 5,979 women concluded that lesbians had a higher prevalence of being overweight and obese than other nonhomosexual women (Boehmer, Bowen, & Bauer, 2007).

WAIST-TO-HIP RATIO IN WOMEN

Related to studies of height and weight is the study of body morphology, shape, or build. At puberty the surge in androgen production

TABLE 10.4. Summary of Weight Studies

Finding	Number of Studies
GayM weigh less than HetM[a]	4
GayM weigh the same as HetM[b]	3
Lesbians weigh more than HetW[c]	4
Lesbians weigh the same as HetW[d]	1

[a]Evans (1972); Blanchard, Dickey, & Jones (1995); Bogaert and Blanchard (1996); Blanchard and Bogaert (1996a).
[b]Yager, Kurtzman, Landsverk, & Wiesmeier (1988); Siever (1994); Bogaert and Friesen (2002).
[c]Perkins (1981); Siever (1994); Bogaert (1998a); Boehmer, Bowen, & Bauer (2007).
[d]Bogaert and Friesen (2002).

in men results in fat deposits primarily in the abdomen, with a resultant waist-to-hip ratio (WHR) of 0.85 to 0.95 (Singh, 1995). In contrast, the pubertal surge of estrogens in women results in fat deposits in the buttocks and thighs and a resultant WHR of 0.67 to 0.80 (Singh, 1993). Thus, as with many other anthropometric measures, WHR is sexually dimorphic. The most recent study examined WHR with respect to sexual orientation, comparing self-described "butch" and "femme" lesbians with heterosexual women (Singh, Vidaurri, Zambarano, & Dabbs Jr., 1999). When compared to each other and a sample of heterosexual women, butch lesbians were found to have a higher WHR, that is, a more masculine build, than either femme lesbians or heterosexual women. No significant difference in the WHR was observed between femme lesbians and heterosexual women. These results confirm those of a previous study by Perkins (1981) that lesbians have narrower hips than heterosexual women. In addition to a higher WHR, Singh et al. (1999) also reported that WHR for lesbians correlated with recalled childhood gender atypical behavior (tomboyness) and adult levels of testosterone. Finally, it was concluded that the evidence suggests that differences in the current hormonal levels by sexual orientation correlate with the degree of masculinity in behavior and body build.

PENIS SIZE

An additional morphological aspect that has been examined in gay men versus heterosexual men is penis size. Two studies have examined penis size with respect to sexual orientation. Nedoma and Freund (1961) studied 126 gay men and 86 heterosexual men and found that gay men had larger penises. The data for this study was obtained by physician measurement of penis size. Bogaert and Hershberger (1999) obtained data on penis size from the original Kinsey sample (Gebhard & Johnson, 1979; Kinsey et al., 1948). The study population consisted of a total of 5,122 predominantly Caucasian individuals, 935 of which were gay men while 4,187 were heterosexual. Data reflect measurements determined by the study subjects. On all 5 measures of penis size, including flaccid and erect length and circumference, gay men reported larger penises than did heterosexual men. The difference between the 2 groups was small, with average percent differences in length being 5.5 percent and in circumference

3.4 percent, but highly significant and not due to differences in height or weight.

ONSET OF PUBERTY

The timing of the onset of puberty, as with other anthropometric measures, is sexually dimorphic; boys reach puberty later than girls (Reiter & Rosenfeld, 1998). It is reasoned that gay men would resemble heterosexual women and reach puberty earlier than heterosexual men because the relevant sexually dimorphic brain structures involved in the timing and regulation of puberty would be more similar to the female-typical brain structures (Blanchard & Bogaert, 1996a). Using similar reasoning, lesbians would be more similar to heterosexual men and reach puberty later than heterosexual women (Bogaert, 1998a, 1998b). Using the physical signs of puberty of first ejaculation and appearance of pubic hair, gay men have been shown to reach puberty earlier than heterosexual men (Kinsey et al., 1948; Blanchard & Bogaert, 1996a; Bogaert & Blanchard, 1996; Bogaert, Friesen, & Klentrou, 2002). Similar results have been found using behavioral signs of puberty (Bogaert & Blanchard, 1996; Bogaert & Friesen, 2002). In all, 6 studies concluded that gay men reach puberty earlier than do heterosexual men. The average age of onset of puberty across the 6 studies for gay men and heterosexual men was respectively 12.62 years and 12.96 years.

On the other hand, using physiological signs of puberty in women, that is, age of first menstrual period, 6 separate studies have found no significant difference in the timing of the onset of puberty between lesbians and heterosexual women (Bell, Weinberg, & Hammersmith, 1981; Tenhula & Bailey, 1998; Bogaert, 1998a, 1998b; Singh et al., 1999; Bogaert & Friesen, 2002; Bogaert et al., 2002). The average age of onset of puberty across the six studies for lesbians and heterosexual women was respectively 12.69 years and 12.69 years. However, the most recent study of the onset of puberty in male and female adolescents does not concur with previous findings (Savin-Williams & Ream, 2006). The study followed adolescents (10,828 participants) over a period of approximately 5 years and assessed measures of puberty 3 times during this period. On most pubertal measures, same-sex groups did not differ from heterosexuals. The exceptions were that homosexual males were more likely to report having a later rather than an earlier onset of puberty relative to heterosexual males, and homosexual females tended to have an earlier onset of puberty relative to heterosexual females.

SUMMARY

Long Bones

Measurement of long bones reveals that on all measures except height, gay men are smaller than heteroseuxal men, while lesbians are larger than heterosexual women. These finding are consistent with the neurohormonal theory.

Finger-Length Ratio

The studies thus far support the conclusion that the 2D:4D finger-length ratio in lesbians is lower than that for heterosexual women and comparable to that reported for females with CAH. The 2D:4D finger-length ratio study of monozygotic twins also found that the ratio for the lesbian twin to be lower than that for her heterosexual twin. These studies support the conclusion that exposure to androgens during gestation influences the 2D:4D ratio and sexual orientation. At present, no conclusions with regard to the 2D:4D ratio in gay men versus heterosexual men is justified.

Fingerprint Analysis

To date a number of studies have compared the fingerprint characteristics of gay men versus heterosexual men. These studies agree that no difference in total ridge count or directional asymmetry of fingerprints exists between gay men and heterosexual men. One study of a small number of monozygotic female twins found the lesbian twin to have lower ridge counts as compared to her heterosexual twin. A separate population study found no difference in total ridge counts between lesbians and heterosexual women. Thus, no definitive conclusions are justified at present.

Height and Weight

At present two out of five studies have found gay men to be shorter than heterosexual men, while four out of seven studies found gay men weigh less that do heterosexual men. Thus, no definitive conclusions with respect to either height or weight differences are justified at present.

One of two large studies of women found lesbians to be significantly taller than heterosexual women. The other study found lesbians to be taller, but the difference did not reach statistical significance. This was particularly true for lesbians who were psychosexually more dominant. Four out of five studies of the weight difference between lesbians and heterosexual women found lesbians to be heavier.

Waist-to-Hip Ratio in Women

When lesbians who self-identify as "butch" and "femme" were compared to each other and to heterosexual women, a significant difference in waist-to-hip ratio (WHR) was found between "butch" lesbians and both "femme" lesbians and heterosexual women. No difference was found between "femme" lesbians and heterosexual women.

Penis Size

Two studies have found that gay men have slightly larger penises than do heterosexual men by on average 5.5 percent in length and 3.4 percent in circumference.

Onset of Puberty

Six studies concluded that gay men reach puberty earlier than do heterosexual men. The average age of onset was 12.62 years for gay men and 12.96 years for heterosexual men. In six studies, no significant differences were found in the age of onset of puberty between lesbians and heterosexual women. However, a recent study of 10,828 participants found no difference between homosexuals and heterosexuals. Thus, the issue for gay men is not clear. Data do support the

TABLE 10.5. Summary of Anthropometric Data

Measure	Lesbians	Gay Men
Long bones	+	+*
2D:4D digit ratio	+	?
Fingerprint analysis	+?	-
Height	+?	?
Weight	+	?
WHR	+**	
Penis size		-***
Age of onset of puberty	-	+

*Except height.
**Women only.
***Hypermasculinized.

conclusion that no difference in the age of onset of puberty exists between lesbians and heterosexual women.

If we characterize those data that show lesbians to be masculinized and gay men to be feminized on a measure as conforming to the neurohormonal theory, we can tabulate the data as follows (see Table 10.5):

The data for lesbians conforms pretty well to the neurohormonal theory, whereas the data for gay men is very mixed at present. On two measures, the data for gay men don't conform at all and on one of these, penis size, gay men appear hypermasculinized. These physical differences, whether they conform to the neurohormonal theory or not, are still likely to be biologically determined.

Chapter 11

Sensory Systems

A number of sensory system functions are sexually dimorphic and have for this reason been studied in gay men and lesbians relative to heterosexual men and women. We will look at hearing responses, the eye-blink startle reflex, and the reaction to smelling specific odors.

HEARING

Otoacoustic Emissions (OAEs)

Otoacoustic emissions (OAEs) are sounds that originate in the inner ear and spread back out into the external ear canal where they can be detected (see McFadden, Loehlin, & Pasanen, 1996; McFadden, 2002). The two types of OAEs that are sexually dimorphic are spontaneous OAEs (SOAEs) and click-evoked OAEs (CEOAEs). Spontaneous OAEs are pure tones that are continuously emitted by most normal-hearing ears. Spontaneous-OAE patterns vary from ear to ear, but the pattern for a given ear is stable across time and can be detected with a small microphone system placed in the ear canal. Click-evoked OAEs are echolike sounds produced by the ear in response to a click stimulus. While they are very individualistic, they are also reasonably stable over time. It is thought that the mechanisms underlying SOAEs and CEOAEs are overlapping but not identical.

Some of the characteristics of OAEs are the following:

Nature's Choice

- SOAEs are pure tones emitted continuously by most normal-hearing ears
- SOAEs are more prevalent in females that males, with 75 to 85 percent of females having at least one SOAE compared to 45 to 65 percent of males.
- CEOAEs are present in all normal-hearing ears, and are stronger in females than in males.
- SOAEs are more prevalent and CEOAEs are stronger in right ears than in left.
- The sex differences seen in OAEs of adults are present in new-borns.
- Hearing sensitivity is better in females than males, and is better in right ears than in left.
- SOAEs appear constant throughout life at least in the frequency regions retaining normal hearing sensitivity.
- SOAEs are generally weak, and most people do not hear their own SOAEs, so they are not the basis for the ringing in the ears (tinnitus) that commonly accompanies hearing loss.
- The heritability of OAEs is about 75 percent; that is, genes contribute significantly to the presence and characteristics of an individuals OAEs.

OAEs in Women From Opposite-Sex Twin Pairs

Studies of women having a male co-twin (dizygotic or nonidentical twins) found that OAEs were displaced toward those of males; that is, their OAEs were reduced and appeared to be masculinized relative to all other types of females tested (McFadden, 1993; McFadden et al., 1996). When the opposite-sex female twins (male-female twins) were compared with same-sex female twins (female-female twins), the difference was significant for SOAEs and marginally significant for CEOAEs. The OAEs of opposite-sex dizygotic male twins (male-female twins) were not different from those of other males.

OAEs in Homosexuals

A separate study examined OAEs in homosexual and heterosexual individuals (60 lesbians and bisexual women, 57 heterosexual women, 51 gay men, 11 bisexual men, and 56 heterosexual men). The differences in OAEs observed between women and men, lesbians and heterosexual women, and gay men and heterosexual men are considered for the most part medium effects. The OAEs of lesbian and bisexual women were shifted towards males; their OAEs appeared masculinized compared to those of heterosexual women (McFadden & Pasanen, 1998, 1999). No difference was found between OAE expression in gay men and heterosexual men.

Supportive Animal Experiments

For technical reasons, click-evoked otoacoustic emissions (CEOAEs) are not easily measured in most animals. However, they have been measured in two mammals, the rhesus monkey (*Macaca mulatta*) and the spotted hyena (*Crocuta crocuta*). As in humans, CEOAEs are stronger in females than in males, and the differences were greater in the fall than in the summer (McFadden, Pasanen, Raper, Lange, & Wallen, 2006). The authors interpreted the seasonal changes in OAEs as a reflection of activational hormonal effects and the basic sex differences in OAEs as a reflection of organizational effects of prenatal androgen exposure. Through experiments conducted by other investigators (see McFadden, Pasanen, Raper, et al., 2006 and references therein), monkeys that had been treated during gestation with either testosterone (testosterone enanthate) or an androgen-receptor blocker (flutamide) were available for CEOAE testing. As expected, prenatal testosterone treatment weakened CEOAEs in females, while prenatal androgen-receptor blocker treatment strengthened CEOAEs in males. The authors concluded that the data from both humans and monkeys suggest that the mechanism of OAE production that underlies CEOAEs is sensitive to prenatal androgen-dependent processes.

In a second study, McFadden and his colleagues measured CEOAEs of male and female spotted hyenas (McFadden, Pasanen, Weldele, Glickman, & Place, 2006). Because the female spotted hyena is exposed to high levels of androgens during gestation (see Chapter 4), it was expected that the difference between CEOAEs in females and males would not be significantly different. The CEOAEs measured in nine females and seven males confirmed this expectation. Once again this supports the idea that the CEOAEs are reduced under the influence of androgens during gestation. The CEOAEs measured in three female and three male hyenas exposed to the androgen-receptor blocker during gestation were stronger than those of normal animals. This further confirms the inverse relationship between gestational androgen levels and stronger CEOAEs, high androgen levels, weak CEOAEs and low androgen levels, strong CEOAEs.

Auditory Evoked Potentials

Auditory evoked potentials (AEP) are brain waves produced in response to click stimuli and recorded using scalp-electrodes, a kind of sound electroencephalogram. Sex and ear differences exist in certain AEP measures in infants and adults (McFadden and Champlin, 2000). A study of AEPs in heterosexual and homosexual individuals (57 lesbian and bisexual women, 49 heterosexual women, 53 gay and bisexual men, and 50 heterosexual men) revealed that selected measures were sexually dimorphic. For these measures (specific amplitudes of selected wave peaks and selected latencies), the values for lesbians were intermediate between those of heterosexual women and heterosexual men and thus appeared masculinized on these measures. In contrast, the measures for gay men were shifted even further from heterosexual women than were heterosexual men and thus appeared to be hypermasculinized on these measures.

In his review of OAEs and AEPs McFadden (2002) identified the following observations: (1) there are sex and ear differences in OAEs and AEPs and hearing sensitivity, (2) these sex and ear differences exist in newborns and adults, (3) OAEs and AEPs appear to be reasonably stable traits throughout life, (4) females from opposite-sex twin pairs have masculinized OAEs, (5) lesbians and bisexual women have masculinized OAEs and AEPs, and (6) gay men have hypermasculinized AEPs, but comparable OAEs relative to heterosexual men.

McFadden (2002) suggests that the finding that OAEs and some AEPs exhibit sex differences and that these sex differences are present in adults, young children, and newborns, implies that these differences are the result of mechanisms operating prenatally and differentially in the two sexes. Furthermore, the differences in OAEs observed in females of opposite-sex twins and in the OAEs and AEPs of heterosexuals and homosexuals exist at birth. If this is true, then it is reasonable to conclude that the sex differences seen in the OAEs and AEPs of newborns is the result of the same mechanism known to be responsible for numerous other sex differences in body, brain, and behavior, namely differential exposure to androgens during prenatal development. This leads to the viewpoint that greater exposure to androgens during prenatal development leads to a weakening of the cochlear amplifiers, which produces a weakening of the OAEs and a small loss in hearing sensitivity and also leads to effects on various auditory nuclei concerned with hearing sensitivity, AEPs, and possibly other auditory characteristics or abilities. This prenatal androgen exposure explanation provides a general overall perspective that appears consistent with all of the auditory results considered in the review.

The author took care to note that the findings he reports are all correlations and do not demonstrate a cause-and-effect relationship. The observed results could easily relate to a causal factor, other than androgen exposure, that was also responsible for the person's membership in the group of interest, that is, males versus females, lesbians versus heterosexual women, female twins of opposite-sex twin pairs versus females of same-sex dizygotic twin pairs, and gay men versus heterosexual men.

McFadden goes further and poses three questions that need to be addressed to identify possible mechanisms by which androgens might exert their effects:

- What are the specific mechanisms responsible for group differences in androgen exposure?
- Why do the groups differ from their comparison groups on certain characteristics and traits but not on the majority or all characteristics and traits?
- Why do gay men differ from heterosexual men in the opposite directions for different characteristics and traits?

In addressing these questions, McFadden (2002) notes that the hypermasculine effect seen for AEPs in gay men are not unique in the homosexuality literature; hypermasculine effects in gay men have been reported for 2D:4D finger-length ratio (Robinson and Manning, 2000; but see Williams et al., 2000), left handedness (Lalumière, Blanchard, & Zucker, 2000; but see Zucker, Beaulieu, Bradley, Grimshaw, & Wilcox, 2001), and penis size (Bogaert & Hershberger, 1999). On many measures, gay men differ from heterosexual men by being intermediate between heterosexual men and heterosexual women, or hypomasculinized. Furthermore, on many measures gay men do not differ from heterosexual men at all. So, how do we explain these observations?

McFadden (2002) identifies two possible factors to explain these findings with respect to gay men and lesbians. First, "localized effects" having the following features:

- Different structures and neural circuits of the body and brain are responsible for the different characteristics, traits, and abilities that have been compared in homosexuals, heterosexuals, and other populations.
- These different structures and circuits can be differentially and possibly independently affected during development.
- The reason for this differential effect could be that the relevant controlling mechanism(s) operate locally in space and time rather than globally.
- Multiple mechanisms may operate on each localized structure or circuit.

Thus, differences between groups in sex-related characteristics and traits may not originate from global, relatively long-lasting prenatal differences in androgen levels, but rather from differences in the androgen concentration in certain localized structures during a specific time period of prenatal development. So, the global androgen levels as measured in intrauterine fluids, maternal and fetal blood, and cerebrospinal fluid may be identical for two individuals who may ultimately have different sexual orientations. They may have had differences in localized androgen levels in the body and/or brain at specific times during gestation. However, the androgen levels for the homosexual individual were higher or lower than is typical for heterosexual individuals.

The second factor, related to the localized androgen effect, concerns the combined effect at a site of the androgen concentration, the number of available androgen receptors, and the rate of conversion of testosterone to estradiol by the enzyme aromatase. Reports suggest that the precise timing of the localized difference in androgen concentration, receptor density, and/or aromatization rate may determine to what degree a localized structure is masculinized (Beach, 1975;

Goy et al., 1988). This concept should not to be interpreted as exhaustive or exclusive of other possible mechanisms. Others have also suggested local rather than global mechanisms of sexual differentiation of the brain (see Woodson and Gorski, 1999).

OAEs, APEs, and Sex and Sexual Orientation Traits

If prenatal androgen exposure is related to masculinization of OAEs and AEPs, then it is reasonable to ask to what degree other sexually dimorphic traits affected by prenatal androgens correlate with the shifts observed for OAEs and AEPs in lesbians and gay men relative to heterosexual control groups. To this end, Loehlin and McFadden (2003) collected an extensive array of data over two separate studies (some through questionnaire and some through direct investigator determination) during the course of studies of OAEs and AEPs. The questionnaire, tests, and rating measures included physical characteristics, spatial abilities, sex roles and sexual orientation, childhood gender nonconformity, and the presence of homosexuality or bisexuality among relatives. The study subjects included 499 college students (106 heterosexual men, 114 homosexual or bisexual men, average mean age of 23.9 years; 149 heterosexual women, 130 homosexual or bisexual women, average mean age 22.8 years).

Although the majority of the questionnaire, test, and rating differences did not show a statistically significant correlation with the OAEs and AEPs, they tended to be consistent with the hypothesis of hypermasculinization of homosexual women and hypomasculination of homosexual men. No support for the linkage of some male homosexuality and the X chromosome was found. The data did support the relationship between childhood gender nonconformity for both homosexual men and women. There was also support for familiality of both male and female homosexuality. The authors caution that these correlative data may suggest a common origin in prenatal androgen effects, or that one might be the consequence of the other. Last, the authors suggest, "it seems unlikely that any very simple prenatal androgenization hypothesis will succeed in explaining the entire pattern of individual and group differences on auditory and non-auditory traits" (Loehlin & McFadden, 2003, p. 125).

EYE-BLINK STARTLE REFLEX

Another type of sexually dimorphic sensory response is prepulse inhibition (PPI). Prepulse inhibition is a startle response and measures individual differences in attentional-information processing. Prepulse inhibition refers to a consistent reduction in the startle response to a strong sensory stimulus (pulse) when it is preceded at an interval of 30 to 500 milliseconds by a weak stimulus, the prepulse (Graham, 1975). Prepulse inhibition is not learned, is not a form of conditioning, and is observed in human infants (Hoffman & Ison, 1992; Swerdlow, Caine, Braff, & Geyer, 1992; Swerdlow, Braff, Taaid, & Geyer, 1994). Strong sex differences exist in humans and rats, with females displaying lower PPI than males in both species (Faraday, O'Donoghue, & Grumberg, 1999; Swerdlow et al., 1993). Prepulse inhibition can be reliably elicited and measured (Braff et al., 1978), and it is very stable in healthy individuals (Cadenhead, Carasso, Swerdlow, Geyer, & Braff, D.L., 1999).

A recent study examined PPI in heterosexual and homosexual individuals (Rahman, Kumari, & Wilson, 2003). In all, 59 subjects participated in the study, 14 lesbians, 15 heterosexual women, 15 gay men, and 15 heterosexual men. Both the pulse and prepulse (sound of lower intensity) are sound-startle stimuli and are presented to the subject through headphones. The monitored responses are eye-blink startle reflexes measured by electromyographic activity of the muscle that causes the eye to blink. The measured amplitude of the muscle contraction for the pulse alone (P) and prepulse plus pulse (PP) are averaged over six trials, and the resultant averages are used to calculate the prepulse inhibition as a percentage;

$$\text{percent inhibition} = \frac{P - PP}{P} \times 100.$$

Significant differences in the percent inhibition were observed between heterosexual men and heterosexual women, with men displaying about a threefold greater inhibition; PPI for heterosexual men was about 40 percent and for heterosexual women about 13 percent. Compared to heterosexual women, lesbians displayed a 2.3-fold greater inhibition; PPI for lesbians was about 33 percent, that is, much closer to the inhibition observed for heterosexual men. The PPI for gay men was not significantly different than that for heterosexual men.

Thus, as was found for OAEs and AEPs, PPI is sexually dimorphic in heterosexual men and women, with men displaying much greater inhibition than women (Faraday et al., 1999: Swerdlow et al., 1993). In addition, lesbian women displayed a masculinized PPI that was statistically no different from that of heterosexual men, as was found for OAEs (McFadden & Pasanen, 1998, 1999). No significant differences in PPI were observed for gay men relative to heterosexual men.

As stated by the authors, PPI represents a basic, nonlearned, cross-species reflex, and the current data suggest that sexual orientation-related differences in PPI reflect differences in early "hardwiring" of the limbic circuitry and some cortical regions underlying the startle response (Rahman, Kumari, et al., 2003). This is consistent with the evidence for masculinized OAEs and 2D:4D finger-length ratios in lesbians, which are known to differentiate in utero under the influence of gonadal steroids, especially androgens (McFadden & Pasanen, 1998, 1999; McFadden & Shubel, 2002; Rahman & Wilson, 2003b; Williams et al., 2000).

That no difference in PPI was found between gay men and heterosexual men was noted by the authors as not being a unique finding. On some measures, gay men appear hypomasculinized, and on others hypermasculinized. The effects of androgens on lesbians appear linear; as the prenatal levels of androgen increase, the degree of masculinization of the individual increases. In contrast, the effects of prenatal androgen on gay men may be nonlinear; both high and low levels of androgens cause deviations from male-typical measures. This has been referred to as "non-monotonic effects" by McFadden (2002). The relationship between prenatal androgen levels and the degree of masculinization in gay men are characterized as follows: increasing androgen exposure produces increasing masculinization up to a point, beyond which further increases in androgen produce a reversal toward the original state. This hypothesis suggests that an overexposure to androgens could produce a change in body, brain, and behavior that is either hypermasculinized, hypomasculinized, or neither, depending on the unique characteristics of the measure under consideration. Examples of this have been observed in animals and reported in the literature.

SMELLING

Pheromones are volatile compounds (molecules) secreted into the environment (in sweat or urine) by one individual of a species, perceived by another individual of the same species, in whom they trigger a behavioral response or physiological change (Karlsen & Luscher, 1959, as cited in Savic, Berglund, Gulyas, & Roland, 2001). In the majority of mammals, pheromones are transduced, or converted into electrical signals that reach the anterior hypothalamus. Pheromones influence sexual behavior and reproductive functions in a sex-specific way via the hypothalamus and its associated connections. It is uncertain how potential pheromone signals work in adult humans, but the question of whether there are compounds that by smelling activate the hypothalamus in humans in a sex-specific mode is of considerable scientific interest.

Savic and colleagues (2001) reported a study of the possible sex differences in brain **activation** by two different sex hormone-like compounds in both males and females. The two compounds were AND (4,16-androstadien-3-one), a testosterone-like compound produced in human underarm secretions in concentrations that are up to 20 times higher in men than in women; and EST (estra-1,3,5(10),16-tetraen-3-ol), a substance similar to naturally occurring estrogens. Brain activation was determined by measuring regional cerebral blood flow (rCBF) with positron emission tomography (**PET scan**). PET scan is a noninvasive imaging technique for visualizing live, real-time changes in cerebral blood flow and metabolism that accompany mental activities. With PET the specific areas of the brain activated by smelling different compounds can be identified. Twelve healthy men and women participated in the study to ask:

- Do AND and EST activate the human brain?
- Do AND and EST activate different regions in males and females?
- Are the activations located in regions of the brain known to mediate sexual and reproductive behavior?

Smelling of AND and EST was assumed to cause brain activation if rCBF was higher than when smelling air. In females AND activated the anterior-ventral hypothalamus, but not the olfactory regions that deal with ordinary odorants, including EST for females. Males, on the other hand, activated the hypothalamus but not the olfactory regions when smelling EST; for males, AND activated the olfactory regions. Thus, the pattern of activation by AND and EST displayed reciprocal features. In further examination of the areas activated by AND and EST in females and males the areas of highest activation were statistically unique. Furthermore, the magnitude of the responses was maximal for AND in females and EST in males. These data lend significant support to the idea that pheromones do function in humans.

A second series of studies by the same group examined brain responses to AND and EST in gay men and lesbian women. In the first study, using a similar experimental approach, 36 healthy subjects (12 each of heterosexual men, gay men, and heterosexual women) were monitored for brain responses to smelling pheromone-type compounds versus air (Savic, Berglund, & Lindstrom, 2005). The questions posed were:

- In gay men, is the hypothalamus activated by AND, EST, or both?
- Is the pattern of activation in gay men similar to that in heterosexual men and heterosexual women, or are entirely different regions activated in gay men?
- If there are group differences, are they confined to the pheromone-like compounds, or do they occur with ordinary odors as well?

Analysis of the data revealed that gay men process AND in the same fashion as heterosexual women rather than as heterosexual men do. Activation for both heterosexual women and gay men was in the preoptic and **ventromedial** hypothalamus when smelling AND, but not EST. Gay men showed no hypothalamic involvement in common with heterosexual men. Ordinary odorants produced the same results in all three groups. The authors conclude that gay men differ from heterosexual men and resemble heterosexual women in that their hypothalamus was activated by AND, with a maximum in the preoptic area. Furthermore, the colocalization of hypothalamic responses with brain circuits that are involved in human reproduction and that in animals are designed to recognize sex further indicates hypothalamic involvement in physiological processes related to sexual orientation in humans.

The second study from this lab asked the same questions as posed previously for gay men, but for lesbians (Berglund, Lindstrom, & Savic, 2006). Using the same experimental design used in their previous studies, brain activation was measured in 12 lesbian women while they smelled AND, EST, four ordinary odors, and air. The results were compared to data obtained from previous studies of heterosexual men and women. Group comparisons showed that lesbians differed only from heterosexual women and that the difference was the absence of preoptic activation with AND in lesbians and its presence in heterosexual women. In addition, the lesbians shared a cluster with heterosexual men in the anterior hypothalamus when smelling EST, but only at a modest significance level. However, when the significance level was lowered slightly, the activated areas in heterosexual men and lesbians became more similar, but the dissimilarity between lesbians and heterosexual women remained the same. If the lesbian women who scored 6 on the Kinsey scale were compared to heterosexual men, the results for EST remained the same, but with a high level of significance. For those specific areas of interest in the hypothalamus, no difference with AND or EST between lesbians and heterosexual men was found. In contrast to AND, EST and ordinary odors showed no significant activation in these same areas. The authors concluded that lesbian women differed from heterosexual women in that they activated the preoptic hypothalamus with AND. In addition, lesbian women shared a hypothalamic cluster with heterosexual men when smelling EST. Last, when focusing on specific areas of the hypothalamus, lesbians showed activation of the dorsomedial and paraventricular hypothalamic area with EST, like heterosexual men and unlike heterosexual women. Overall, these data suggest that lesbians process AND and EST more similar to heterosexual men than to heterosexual women.

These PET scan studies of brain activation while smelling prospective pheromones are very intriguing. What we would like to know is what are the physiological consequences of these activations? In a recent study, 21 healthy heterosexual women were studied after smelling androstadienone (4,16-androstadien-3-one) (AND) and a control substance (CONT) (Wyart et al., 2007). Eight autonomic system measures were monitored continuously during each experimental session, including skin conductance, electrocardiogram, pulse (finger and ear), blood pressure, skin temperature, abdominal respiration, and thoracic respiration. In addition, each subject was monitored for body movement or fidgeting and mood was assessed by means of a 17-item scale. Saliva samples were collected for the determination of cortisol levels.

Smelling AND increased positive mood and sexual arousal relative to CONT, consistent with previous studies (Wyart et al., 2007, and references cited therein). Also, salivary cortisol significantly increased after smelling AND as compared to CONT. Finally, the entire experimental protocol was repeated with an additional 27 subjects, with nearly identical results. Thus, AND may qualify as a human pheromone. The authors note the key contribution of their findings is the identification of a single molecule capable of triggering an endocrine response. The authors also caution that there may be many more molecules in sweat that can induce a variety of endocrine responses. What these data also provide is at least one of the endocrine correlative responses to the brain activity findings discussed previously. How they relate will require a great deal more work.

SUMMARY

The auditory studies of lesbians demonstrate once again that with respect to otoacoustic emissions and auditory evoked potentials lesbians appear masculinized relative to heterosexual women. These observations are supported by animal experiments that demonstrate a clear inverse relationship between prenatal testosterone and the strength of OAEs. Similarly, lesbians appear masculinized with regard to the eye-blink prepulse inhibition response. Furthermore, since there is evidence that these responses develop during gestation and appear dependent on androgens, these observations are consistent with the neurohormonal theory.

The auditory responses of gay men were mixed. The auditory evoked potentials of gay men were hypermasculinized relative to heterosexual men. No significant difference in otoacoustic emissions or eye-blink prepulse inhibition was observed between gay men and heterosexual men. These findings suggest that the prenatal effects of androgen on the sexual differentiation of the auditory system in gay men are not simply linear and are likely more complex than observed for lesbians.

Support for pheromone function in humans has emerged, with activation of the hypothalamus associated with hormone-like compounds found in male and female sweat and urine. As anticipated, heterosexual men respond to a compound produced by women and heterosexual women respond to a compound produced by men. The response of gay men was similar to that of heterosexual women, while the response of lesbians was similar to that of heterosexual men. Thus, once again, the response of homosexual individuals appears gender inverted relative to that of heterosexual individuals and is consistent with the neurohormonal theory.

Chapter 12

Cerebral Lateralization

Another functional trait on which men differ from women is cerebral lateralization, or the degree to which selected motor and cognitive functions are associated with one hemisphere or the other. In general, men are more lateralized, that is, functions are more restricted to one hemisphere. In women, selected functions are more distributed across both hemispheres.

HANDEDNESS

Handedness, which is whether one is right- or left-handed, is another human trait that is sexually dimorphic. A number of large studies of both children and adults have shown that males are somewhat more likely to be left-handed than are females (see Lalumière et al., 2000; Lippa, 2003b). Left-handedness in the general population for men has been reported at 8.5 to 10.6 percent and for women 6.7 to 8.5 percent; the values for children are 11.8 percent for boys and 9.0 percent for girls. The reported incidence of left-handedness for men is 27 percent higher than that for women (for a review see Lippa, 2003b). Although the biological basis for the difference in handedness between men and women is not understood, several studies suggest that a number of masculine characteristics are associated with left-handedness in women (see Lippa, 2003b). Evidence that shifts in gender-related traits are associated with left-handedness in men is much more limited.

Studies of human fetuses, neonates (infants during the first month of life), and infants have shown a right-hand preference that is stable over time and closely matches the level of right-handedness in the

Nature's Choice

general population. Thus, handedness is determined early in development and likely well before birth. Handedness is also associated with
cerebral laterality, or which sides of the brain support different functions, such as language or visual-spatial abilities. Smaller numbers of
left-handers display the cerebral laterality displayed by right-handers.
Left-handers are subject to a number of neurodevelopmental problems, which are also more prevalent in men.

Because handedness is sexually dimorphic, determined early in
development, and related to brain development, it has also been studied with respect to sexual orientation. The expectation based on the
neurohormonal theory is that handedness in gay men will be feminized, more similar to handedness of heterosexual women than of
heterosexual men, and that handedness of lesbians will be more
masculinized, more similar to heterosexual men than heterosexual
women.

The most comprehensive study published to date is a meta-analysis of 20 previously published studies of handedness in homosexual
and heterosexual men and women (Lalumière et al., 2000). Selection
criteria for inclusion of studies in the meta-analysis were (1) that both
a heterosexual and homosexual population were assessed and (2) that
a well-specified measure of handedness was used that was the same
for both populations and (3) that the data were broken down by sex.
In all, handedness was compared in 6,987 homosexual individuals
(6,182 men and 805 women) and 16,423 heterosexual individuals
(14,808 men and 1,615 women). The analysis revealed that gay men
had a 34 percent greater odds of being left-handed relative to heterosexual men, while lesbians had a 91 percent greater odds of being
left-handed relative to heterosexual women. The authors concluded
that their results support the view that sexual orientation in some men
and women has an early neurodevelopmental basis, but the factors responsible for the association between sexual orientation and left-
handedness have yet to be determined.

Two additional studies have been published on the relationship between handedness and sexual orientation. The first study examined
handedness in 382 men (177 heterosexual and 205 gay men) and 354
women (205 heterosexual and 149 lesbian women) (Mustanski, Bailey,
et al., 2002). Heterosexual participants were undergraduate students
and homosexual participants were attendees of a transgender conference, various gay and lesbian college groups, and a gay men's chorus.

Sexual orientation was assessed using the Kinsey Scales. Handedness was assessed by self-report via a questionnaire. The study found that left-handedness in gay men did not differ significantly from that for heterosexual men; left-handedness for each was 10 percent of the populations examined. In contrast, lesbians displayed greater left-handedness than did heterosexual women, which was primarily due to the relatively high percentage of lesbians who were ambidextrous. The combined ambidextrous and left-handedness for lesbians was 16 percent, while that for heterosexual women was 10 percent. The authors concluded that the existence of an association between left-handedness and sexual orientation does provide more suggestive evidence for the view that sexual orientation is at least partly inborn, that is, biological in origin.

The second study examined handedness in 350 heterosexual men, 458 gay men, 706 heterosexual women, and 468 lesbians (Lippa, 2003b). Heterosexual participants were college students and university staff members. The homosexual participants were attendees of a gay pride festival, college students, and university staff. Sexual orientation was assessed by self-report. Handedness was assessed by self-report using a 5-point scale, which included: exclusively use right, mostly use right, use both equally, mostly use left, and exclusively use left. A significant difference in handedness was observed between heterosexual men and gay men, with 11.4 percent of heterosexual men and 19.0 percent of gay men classified as non-right-handed. Gay men displayed an increased odds of being non-right-handed of 82 percent relative to heterosexual men. The corresponding data for women revealed a small difference in non-right-handedness that was not statistically significant; 12.0 percent for heterosexual women and 14.3 percent for lesbians. Lesbians had an increased odds of being non-right-handed of only 22 percent. Finally, the combined data for men and women indicated a significant difference in handedness, with 11.8 percent of heterosexuals and 16.7 percent of homosexuals classified as non-right-handed. Homosexuals had an increased odds of being non-right-handed of 50 percent.

The complex relationship between handedness and sexual orientation in men versus women is revealed by breaking the data down across a 5-point scale for handedness. When this was done, there was a significant difference in handedness percentages across the 5 handedness categories for heterosexual and gay men. Similarly, for women, and in contrast to the data presented previously, there was a significant difference in handedness percentages across the 5 handedness categories for heterosexual women and lesbians. The author noted that the results for both men and women indicated that for homosexual individuals, handedness was shifted away from the complete right-handedness. Furthermore, gay men and lesbians seem to show somewhat different patterns of leftward shift. Overall, larger heterosexual-homosexual handedness differences were observed for men than for women.

The odds ratio is the increased odds of being non-right-handed for the homosexual participants compared with the heterosexual participants. This is a useful way of presenting the data, as the odds ratio is independent of the base rates of handedness. A value larger than 1.0 indicates a larger proportion of non-right-handers in the homosexual sample than in the heterosexual sample. Another advantage of the odds ratio is that it provides an easily understandable measure of effect size. For example, a value of 2.00 means that a randomly chosen homosexual participant is twice as likely to be non-right-handed as a randomly chosen heterosexual participant. There is at present no consensus of the data for handedness in gay men and lesbians. We would expect the meta-analysis data of Lalumière et al. (2000) to be the most accurate, as it reflects the combined results of 20 separate studies of men and 9 separate studies of women. The subsequent studies of Mustanski, Bailey, et al. (2002) and Lippa (2003b) not only do not agree, but also do not agree with the meta-analysis data. At present it seems reasonable to conclude that non-right-handedness is increased among gay men and lesbians, but the exact magnitude of the effect is yet to be accurately defined. Also, whether non-right-handedness is greater in gay men or lesbians is also not definitively determined. Furthermore, the refined study of handedness based on a five-point scale of handedness in gay men and lesbians by Lippa (2003b) indicated that handedness was shifted away from complete right-handedness, and that the pattern of leftward shift was different for gay men and lesbians.

Looking only at the odds ratios for lesbians, we can conclude that more lesbians are non-right-handed than are heterosexual women. This is consistent with the result predicted by the neurohormonal hypothesis. In contrast however, the odds ratios for gay men are not consistent with the result predicted by the neurohormonal hypothesis. Thus, we must be open to other theoretical explanations for the relationship between non-right-handedness and homosexuality in men.

COGNITIVE FUNCTIONS

Mammalian brains have a left and a right hemisphere. Communication between the left and right hemispheres in the human brain takes place through crossing nerve fibers in three commissures, or structures containing only nerve fibers that connect the hemispheres; the major connecting structure is the **corpus callosum** and the two smaller structures are the anterior and posterior commissures.

Many neural pathways of the central nervous system are bilaterally symmetrical; that is, they are nearly identical structurally on both sides of the body. In addition, neural pathways cross over to the opposite (contralateral) side of the body, usually in the brain stem or very close to the brain; this includes motor, sensory, auditory, and visual systems. This means that voluntary movements and sensory events are controlled and processed respectively by the cerebral hemisphere on the opposite side of the body. The right hemisphere controls the left side of the body, including hearing and vision, and the left hemisphere controls the right side of the body.

For the majority of people who are right-handed, the processing of incoming sensory information and the control of motor functions are associated with the left hemisphere. In addition, for most right-handed people, language functions are associated with the left hemisphere. Furthermore, most right-handed people are right-ear dominant and right-ear functions are associated with the left hemisphere.

Although the hemispheres look nearly identical when viewing the surface of the brain, they are not really structurally or functionally identical, and the observed asymmetries are not the same in all individuals. For example, the planum temporale is a region on the upper surface of the temporal lobe, which is located on the side of the brain, and is known to be associated with language function. The left planum temporale is larger in 65 percent of human brains, while the right is larger in only 11 percent of brains, and in 24 percent of brains the left and right side are nearly the same size (Kandel, Schwartz, & Jessell, 1995, p. 358). Functional asymmetries also exist. For example, most of the general population is right-handed, which means that the left hemisphere is dominant for hand function, while about 8 to 12 percent of the general population is left-handed (Annett, 1985). Within each handedness group, the extent to which individuals depend on the dominant hand is variable. Individuals also display considerable variability in the lateralization of cognitive functions. While 95 percent of right-handers have left hemisphere language, 61 percent of left-handers have left hemisphere language, 20 percent have language functions distributed to both hemispheres and 19 percent have right hemisphere language (see Grimshaw, Bryden, & Finegan, 1995). Similar differences across individuals are also observed in the lateralization of right hemisphere functions. Studies of patients with localized brain damage demonstrate that selected cognitive functions are preferentially localized to one hemisphere. The left hemisphere excels at sequential, intellectual, analytical, rational, verbal thinking, and motor functions, while the right hemisphere excels at sensory discrimination and visual-spatial, emotional, intuitive, abstract, and nonverbal thinking. Thus, the left hemisphere processes functions related to language and the right hemisphere processes functions related to spatial abilities. Interestingly, aspects of this functional asymmetry are apparent soon after birth, if not before, supporting the idea that factors that contribute to individual differences in cerebral lateralization operate early in development (Witelson, 1987). Furthermore, genetics are thought to play a significant role in an individual's pattern of cerebral lateralization (McManus & Bryden, 1992).

There is evidence that cognitive abilities are sexually dimorphic; on average, women score higher than men on verbal tasks involving verbal association and fluency, perceptual speed, mathematical calculation, and fine manual dexterity, while men score higher than women on mathematical reasoning, visual-spatial tasks, and targeting (Neave, Menaged, & Weightman, 1999). This division of labor between the two hemispheres is referred to as functional cerebral asymmetry (FCA). It can be measured either by recording the electrical activity of the brain from the surface of the head, or by more sophisticated computer-aided imaging techniques, such as function magnetic resonance imaging (fMRI), while a specific task is being performed. It can also be measured by cognitive psychological testing. By these measures, men are characterized as being more lateralized than women. This means that functions are more localized to one hemisphere in men relative to women, for whom functions tend to be more distributed across both hemispheres.

TESTOSTERONE AND CEREBRAL LATERALIZATION

Animal Studies

Animal studies have shown that exposure to androgens during critical periods of brain development lead to masculinization of neuroanatomy, physiology, and behavior (Goy & McEwen, 1980). Furthermore, in the male Long-Evans rat, the right cortex has been found to be thicker than the left, while in the female the reverse is true (Diamond, Johnson, & Ingham, 1975). The female pattern of cerebral asymmetry can be induced in males by castration at birth (Diamond, 1984). On the other hand, testosterone treatment of female rats masculinizes the sexually dimorphic nucleus of the preoptic area of the brain (Arnold & Gorski, 1984), adult sexual behavior (Harris & Levine, 1962), and spatial ability (Williams & Meck, 1991). Prenatal androgen manipulation in rats also affects certain aspects of spatial memory performance. Castration of male rats at birth, which eliminates exposure to testosterone, decreases their performance on spatial memory tasks, resulting in scores in the female-typical range (Williams, Barnett, & Meck, 1990). Female rats treated with estradiol, to which testosterone is converted by neuronal aromatase, outperformed control females and performed more similar to control males on spatial memory tasks. Furthermore, the male advantage in spatial memory tasks can be reversed by treatment with testosterone soon after birth, resulting in better performance by females, but worse performance by males (Roof, 1993). These data suggest that the organizational effect of androgens on spatial behavior is different in males and females. In males, levels of testosterone above or below an optimum for typical male development produce female-typical performance, while in females the relationship between testosterone and spatial ability appears linear; that is, exposure to increasing levels of testosterone results in increasing spatial ability. In primates, testosterone treatment has also been shown to masculinize play behavior (Goy & Phoenix, 1971). Furthermore, it has been suggested that such hormonal manipulations in primates, including humans, could also influence selected aspects of cognitive performance, primarily spatial memory (Williams et al., 1990). In two studies, a relationship between testosterone levels and spatial ability has been demonstrated for men (Silverman, Kastuk, Choi, & Phillips, 1999; O'Connor, Archer, Hair, & Wu, 2001). Data from one study suggests the existence of an optimal testosterone range in men that activates different cognitive functions in a curvilinear fashion. In such a model, suboptimal performance on spatial tests would result in men when their testosterone levels were either above or below the optimal range (O'Connor et al., 2001).

Dizygotic Twins

Dizygotic, or nonidentical, twins are one normal human population that can be studied to assess the effects of testosterone on the development of cognitive functions in females. Spatial abilities were assessed in female members of opposite-sex (OS) and same-sex (SS) dizygotic twin pairs (Cole-Harding, Morstad, & Wilson, 1988). Remember that males score significantly better than do females on spatial tests. Thus, the study asks what the effect of having a male twin in utero on the female's later spatial ability. The OS females' spatial scores were significantly higher than those of the SS females. After three test trials, the scores of the OS females were not significantly different from those of their twin brothers. The scores of the OS males were also not significantly different from those of other male groups. The authors suggested that their results support the possibility that exposure to testosterone in utero improves spatial ability in females, further supporting the theory that differences in prenatal exposure to testosterone are at least partially responsible for the gender difference in spatial ability.

Females with Congenital Adrenal Hyperplasia

The effects of variations in prenatal androgen exposure on humans can also be studied in selected clinical populations that arise due to endocrine disorders, or experiments of nature. Congenital adrenal hyperplasia (CAH) is a condition in which the fetus is exposed to very high levels of androgens beginning in the third month of gestation. Cerebral lateralization in CAH individuals can be assessed by determining handedness. Females with CAH displayed a left-hand preference as compared to their normal sisters, while the hand preference of male patients was not different from their siblings (Nass et al., 1987). This shift in lateralization as indicated by hand preference is interpreted to reflect enhanced development of the right hemisphere relative to the left hemisphere. Two additional studies of handedness in CAH patients have been conducted, with mixed results; one study confirmed the results of Nass et al. (1987) with respect to females, but not for males (Kelso, Nicholls, & Warne, 1999), while the second study found no effect of CAH on handedness (Helleday, Siwers, Ritzen, & Hughdahl, 1994).

Hampson and colleagues studied spatial reasoning, a task that men perform better than women, in a group of 12 individuals between the ages of 8 and 12 and 10 control subjects of comparable age (Hampson, Rovet, & Altmann, 1998). Tests to assess spatial reasoning, a left-brain function, and perceptual speed were administered. Among the controls, boys outperformed girls on the spatial test confirming the results of the meta-analysis of spatial abilities literature by Voyer, Voyer, and Bryden (1995), which showed significant differences in spatial relation scores between boys and girls. This sex difference in spatial reasoning can be reliably observed by age 10. Girls with CAH achieved significantly higher spatial reasoning scores than did control girls and more closely resembled boys in this regard. Overall control subjects displayed stronger perceptual speed ability than did CAH subjects. Furthermore, these finding were not related to general IQ scores.

A second study by Kelso, Nicholls, Warne, & Zacharin (2000) examined cerebral lateralization in a group of 17 individuals with CAH by assessing handedness, verbal IQ, performance IQ (a visual-spatial task), and ear dominance. Subjects with CAH displayed a left-hand bias as compared to their age-matched controls. No gender difference in handedness was observed. Subjects with CAH had significantly higher performance IQ scores than did their controls, while the controls had significantly higher verbal IQ scores than did subjects with CAH. Overall IQ did not differ between CAH and control subjects. Both CAH subjects and controls displayed a right ear hearing advantage. The authors conclude that their results provide support for the hypothesis that elevated prenatal androgen exposure in individuals with CAH affects later patterns of cognitive development and hand preference that reflect a shift in cerebral lateralization toward the right hemisphere. But see Hines (2000) for a discussion of earlier studies that failed to detect differences in cognitive performance of CAH individuals.

Females Exposed to Diethylstilbestrol

In two separate studies cited by Hines (2000), no evidence of differences in cognitive abilities was observed for women exposed prenatally to diethylstilbestrol (DES) relative to their unexposed sisters (controls). Furthermore, a recent epidemiological study of 3,946 women prenatally exposed to DES and 1,749 control women found no difference in handedness between DES–exposed subjects and controls; 11 percent of each group were found to be left-handed, which is the same level of left-handedness found in the general population (Titus-Ernstoff et al., 2003). This result suggests that lateralization of function is comparable between the two groups.

Males with Complete Androgen Insensitivity

Males with complete androgen insensitivity develop in the complete absence of the effects of androgens and thus develop physically and behaviorally as females. A group of 10 individuals with complete androgen insensitivity have been reported to display deficiencies on a number of spatial tasks (Imperalto-McGinley, Pichardo, Gautier, Voyer, & Bryden, 1991).

Turner's Syndrome

In Turner's syndrome (TS), females develop in the complete absence of any ovarian hormones, including testosterone. A study of TS women shows that they have normal verbal IQ, but very significantly depressed abilities on a range of visual-spatial tasks, including those associated with the right hemisphere for which men outperform women (Levy & Heller, 1992).

Thus, we see that functions are distributed differently between the left and right hemispheres in the brains of men and women. Functions in men are more lateralized, that is confined more to one hemisphere, while in women functions are more distributed between both hemispheres. Studies in animals and human dizygotic twins and experiments of nature in humans support the hypothesis that exposure to androgens, and, most important, to testosterone, during gestation is responsible, at least in part if not solely, for these lateralization differences. If this is true, and if the neurohormonal theory of sexual orientation determination is also true, then we can ask, will gay men and lesbians also show some degree of sex-atypical cerebral lateralization? Based on the hormonal gestational theory, we would once again anticipate the brains of gay men to not be as lateralized as those of heterosexual men. Thus, gay men would not be expected to score as well on spatial tests, but to score better on verbal tests than heterosexual men. Furthermore, gay men might score in the same range on both spatial and verbal tests as heterosexual women. In contrast, we might expect the brains of lesbians to be more lateralized than those of heterosexual women. Thus, lesbians should score better on spatial tests, but not as well on verbal tests than heterosexual women; that is, their scores should be shifted toward heterosexual men's scores.

SEXUAL ORIENTATION

Consistent with the predictions based on the neurohormonal theory, no fewer than six early studies revealed that gay men perform in the heterosexual female-typical range on cognitive tests for mental rotation, spatial perception, and measures of verbal fluency (see Rahman, Wilson, & Abrahams, 2004b). The most consistent effects for gay men were found with respect to mental rotation, for which gay men were found to perform poorer than heterosexual men and comparable to heterosexual women (Wegesin, 1998a; Neave et al., 1999). One study of verbal fluency found that heterosexual women outperformed heterosexual men; gay men, on the other hand, scored comparably to heterosexual women on tests of verbal fluency (Wegesin, 1998a). This finding was not confirmed in a later study, however (Neave et al., 1999). These neuropsychological results are consistent with electroencephalographic (EEG) findings in which hemispheric brain activity of gay men was different from heterosexual men, but comparable to that of heterosexual women (Alexander & Sufka, 1993), and with event-related potential results, in which slow wave activity recorded during mental rotation was greater for heterosexual men than for heterosexual women and gay men (Wegesin, 1998a). One report failed to find a sexual orientation related difference for either mental rotation or spatial perception (Gladue & Bailey, 1995). In contrast to the expected result for lesbians based on the neurohormonal theory, these studies found no difference in cognitive functions between lesbians and heterosexual women.

Spatial and Object Location Memory

A recent series of studies by Rahman and colleagues has examined a number of cognitive functions in a large group of both homosexual and heterosexual men and women. The first of these examined spatial memory, which involves the ability to encode, store, and retrieve information regarding route navigation and object location (Rahman, Wilson, & Abrahams, 2003). Object location memory is a spatial ability at which females outperform males. Study participants consisted of 60 heterosexual men, 60 gay men, 60 heterosexual women, and 60 lesbians, all of whom were right-handed. Sexual orientation was assessed using a modified Kinsey Scale. Subjects completed neuropsychological tests to assess (1) object recall, (2) object recognition, and (3) spatial location memory. With regard to object recall, females recalled significantly more objects than males overall, but no effect of sexual orientation was observed. No significant effect of either sex or sexual orientation was observed for object recognition. With regard to spatial location memory, however, heterosexual women performed significantly better than heterosexual males. Gay men also performed significantly better than heterosexual men, but no differently than heterosexual women. The difference between lesbians and heterosexual women was not statistically significant, but lesbians tended to perform more similar to heterosexual men.

Mental Rotation and Judgment of Line Orientation

A second study by Rahman and Wilson (2003a) examined performance in the same groups of heterosexual and homosexual subjects on mental rotation and judgment of line orientation tasks. Large, statistically significant differences in performance on mental rotation tests have been well documented, with men outperforming women (Voyer et al., 1995). Previous reports found that gay men perform comparable to heterosexual women on tests of mental rotation. Lesbians tended to perform more similar to heterosexual men with respect to mental rotation, although the differences were not statistically significant (Wegesin, 1998a). Rahman and Wilson (2003a) found that with respect to judgment of line orientation and mental rotation, men outperformed women. Heterosexual men scored significantly higher than gay men, however gay men did not differ in their scores from heterosexual women. The scores of lesbians did not differ from those of heterosexual women. The same results with respect to mental rotation was observed for men versus women and gay men versus heterosexual men; men outperformed women and gay men, and no difference was observed in the performance of gay men relative to heterosexual women. In contrast, with respect to mental rotation, lesbians outperformed heterosexual women when the data reflected total correct answers. If the data were adjusted to reflect the percent correct answers, lesbians performed comparably to heterosexual women. The authors attribute this difference to lesbians compromising accuracy for speed. Thus, with respect to mental rotation, lesbians are faster than heterosexual women, but no more accurate than heterosexual women. The authors further comment that since a dominant role for the parietal cortex of the brain has been demonstrated by lesion and functional neuroimaging studies these results suggest that differences exist in the functioning of the parietal cortex brain regions between gay men and heterosexual men (Rahman & Wilson, 2003a).

Perceptual Speed

In a third study, perceptual speed was assessed in the same group of subjects using the digit-symbol subtest of the Wechsler Adult Intelligence Scale Revised (WAIS-R) (Rahman, Wilson, & Abrahams, 2004b). This is a test on which women outperform men. Scaled scores for the test revealed that heterosexual women and gay men outperformed heterosexual men, while lesbians performed comparably to heterosexual women. Since this test relies on the executive function capability of the brain, the authors conclude that these results suggest differences in brain executive functioning between gay men and heterosexual men.

Prenatal Hormones and Neurocognitive Functions

In a final study of the same 240 right-handed participants, Rahman and colleagues sought to determine if the differences in neurocognitive functions observed between homosexual and heterosexual men and women was primarily due to the actions of prenatal factors early in development as suggested by the neurohormonal theory, or the influence of social environment, specifically gender role learning (Rahman, Wilson, & Abrahams, 2004a). Gender role learning is characterized as internalized stereotypical notions of male and female personality derived from family, social, and cultural environments that result in the engagement in sex-differentiated activities and behavior by an individual. The neurocognitive performance results obtained as described previously included mental rotation and judgment of line orientation, tasks on which men outperform women, and verbal fluency, perceptual speed, and object location memory, tasks on which women outperform men. Participants were also assessed with respect to (1) psychological gender, how the participants viewed themselves on a "masculinity–femininity" continuum; (2) birth order, of special interest, how many older brothers a male participant had, since the birth order effect demonstrates that the more older brothers a man has, the more likely he is to be gay; (3) sibling sex composition, which is how many brothers and sisters a participant had; and (4) the 2nd to 4th finger-length ratios (2D:4D ratio) of both right and left hands, which is sexually dimorphic and generally sex-atypical for gay men and lesbians, although exceptions have been observed (see Chapter 10). In addition, general intelligence (IQ) and the degree to which participants experienced stress during the course of testing were also assessed. The statistical procedure of stepwise multiple regression analysis was used to determine which of the additionally assessed characteristics predicted the observed performance of each group on the neurocognitive tasks.

With regard to total correct score for mental rotation, sexual orientation was the best predictor of performance, followed by psychological gender as the second best predictor of performance. None of the other characteristics were predictive of mental rotation performance. When mental rotation performance was scored as the percentage correct, sexual orientation and IQ were the strongest predictors of performance.

With regard to performance on the judgment of line orientation task, sexual orientation was the strongest predictor of performance, with IQ a modest second predictor. None of the other characteristics were significantly predictive of performance.

Verbal fluency was assessed using three measures: letter fluency, category fluency, and synonym fluency. The major predictor for all three verbal fluency measures was sexual orientation, followed by age and IQ making comparatively weak contributions. With regard to perceptual speed, three characteristics—sexual orientation, IQ, and psychological gender—were comparably predictive of perceptual speed performance. Last, sexual orientation was the only predictor of object location memory performance.

Strikingly, there was no contribution of 2D:4D ratio, birth order, or sibling sex composition to cognitive performance separated from the other factors, IQ, sex, sexual orientation, and psychological gender. Furthermore, there was also no association between the onset of puberty and cognitive performance for either heterosexual or gay men. Finally, there was also no association between the "degree of butchness" and cognitive performance for either lesbians or gay men.

The authors conclude that their results provide little support for the influence of prenatal hormone exposure on sexually dimorphic cognitive performance. As the authors note, this is in contrast to the findings of others reporting a negative relationship between 2D:4D ratios and mental rotation ability (Rahman et al., 2004a). Psychological gender, an individual's subjective sense of their masculinity or femininity, was a predictor of mental rotation score (total correct). This finding was considered partially consistent with the finding of a limited support for the idea that gender role socialization mediates sex difference in spatial ability, sex having the greater influence (Saucier, McCreary, & Saxberg, 2002). Sexual orientation was by far the most powerful predictor of cognitive differences across those evaluated, with psychological gender and IQ making a small contribution to mental rotation ability.

Navigational Strategy

Spatial navigation is another strongly sexually dimorphic human trait, with women using a landmark strategy on average more than men, and men using an orientation strategy on average more than women (see Rahman, Andersson, & Govier, 2005 and references cited therein). This study assessed navigation strategy in 80 subjects (20 heterosexual men, 20 heterosexual women, 20 gay men, and 20 lesbians) using modified maps (Dabbs, Chang, Strong, & Milun, 1998, from Choi & Silverman, 1996). A number of other cognitive tests were administered at the same time to aid in data analysis. Analysis of the data revealed the following:

- On vocabulary scores, men scored higher than women overall, with heterosexual men scoring significantly higher than heterosexual women. Gay men and heterosexual men did not differ significantly, while lesbians scored higher than heterosexual women.
- On mental rotations, men scored higher than women overall.
- On the road map tests, no significant differences were found other than homosexuals made more errors than heterosexuals.
- On navigational strategy, heterosexual women used significantly more landmarks than heterosexual men, as did gay men. There were no differences between heterosexual women and lesbians in landmark usage.

Overall, men used orientation-based navigational strategy more than women, who used landmarks and left-right directions. Gay men displayed a strong cross-sex shift, using more landmarks compared to heterosexual men and performing comparably to heterosexual women. There were no differences in navigational strategies between lesbians and heterosexual women.

The authors conclude that their study reinforces an increasing body of work that demonstrates strong cross-sex shifts in the neurocognitive performance of gay men, while lesbians do not demonstrate such shifts (see the following section). It should be noted, however, that there is often a trend for lesbians to perform more similar to heterosexual men. Last, the authors point out that it appears that homosexuals of both sexes possess a mosaic-like profile of male-typical and female-typical neurocognitive traits.

Function Cerebral Asymmetry

Functional cerebral asymmetry is, in simple terms, the degree to which a function is more localized to one hemisphere than the other. Data suggest a small but consistent finding that men are more strongly lateralized with respect to verbal information than are women (see Rahman, Cockburn, & Govier, 2007 and references cited therein). Technically, there are different ways to assess lateralization of language function. Dichotic listening is one technique that involves listening to verbal consonant-vowel syllable pairs that elicit a consistent right-ear advantage (REA) indicating greater left hemisphere processing of verbal stimuli in approximately 80 percent of right-handed and 65 percent of left-handed subjects. Interestingly, dichotic listening performance appears related to psychological gender as well as sex. Both males and females in male-typical jobs have greater REA, while men and women in female-typical jobs display reduced REA.

With regard to sexual orientation, increased lateralization for verbal functions was related to increased spatial performance in heterosexual men, but not gay men or women (Wegesin, 1998b). The origins of sexual variation in lateralization are generally attributed to testosterone acting early in development. Language lateralization was studied using the dichotic listening task in 27 heterosexual women, 23 lesbians, and 24 heterosexual men. Heterosexual men scored higher on each laterality index relative to heterosexual women. However, no significant differences were found between heterosexual men and lesbians, or between lesbians and heterosexual women. The same relationship was observed for right-ear scores; heterosexual men displayed significantly greater right-ear scores than heterosexual women. Furthermore, no significant differences were found between heterosexual men and lesbians or between lesbians and heterosexual women. That is, lesbians scored intermediate to heterosexual men and women. The authors conclude that these data show a cross-sex shift among lesbians, in the male direction, using the dichotic listening measure that may reflect prenatal hormone exposure. The authors also note that this is in contrast to a previous report (Rahman et al., 2007).

2D:4D Ratio, Visual-Spatial and Mental Rotation Performance

Using a comparable Internet-based experimental design and drawing on a very large number of study subjects, the relationship of visual-spatial performance (Collaer, Reimers, & Manning, 2007) and mental rotation performance (Peters, Manning, & Reimers, 2007) to sex, sexual orientation, 2D:4D digit ratio, and a number of other variables were examined. The visual-spatial performance study included 116,141 females and 129,482 male. Of those that responded to the question of sexual orientation, 90.3 percent were heterosexual, 4.2 percent were homosexual, and 5.5 percent were bisexual. The respondents completed questionnaires and visual-spatial tests online. Performance and questionnaire results were subjected to analysis of variance. With respect to the visual-spatial task, adult men outscored adult women, and boys (8 to 10 years of age) outscored girls (same age range). With regard to sexual orientation, heterosexual men scored higher than gay/bisexual men, and lesbians/bisexual women scored higher than heterosexual women. When analyzing for homosexuals and heterosexuals as a group, men outperformed women in both the heterosexual and the homosexual/bisexual groups. With regard to 2D:4D ratio and visual-spatial task performance, 2D:4D varied with sex, with men having lower ratios than women. The 2D:4D ratio (right hand) also varied with visual-spatial task performance, with the highest (female-typical) 2D:4D ratios associated with the lowest performance scores, and the lowest (male-typical) 2D:4D ratios associated with the highest performance scores. Since sexual orientation and 2D:4D ratio are thought to be influenced by prenatal androgen exposure, a relationship between them might be anticipated from an analysis of these data. For men, the 2D:4D ratio varied with sexual orientation, with higher digit ratios in gay/bisexual men than in heterosexual men. However, in women, the 2D:4D ratio did not vary significantly as a function of sexual orientation for either the right or left hand.

The authors (Collaer, Reimers, & Manning, 2007) concluded that sex-typed development likely reflects a composite or mosaic of cross-sex and sex-typical behaviors and characteristics, rather than a simple unidirectional shift of all traits. Such a mosaic may result because different traits have different critical periods or steroid sensitivities. This study is complex, and it raises a number of issues that are beyond the scope of this presentation. Suffice it to say that the complexity of the relationships under consideration will require a great deal more work before a clearer understanding is achieved.

A second Internet study examined the relationship between a mental rotation task (MRT) and a number of variables, including height, age, education level, sexual orientation, birth control pill use, and 2D:4D digit ratio (Peters et al., 2007). Subjects for the MRT included 134,317 men and 120,783 women from 7 ethnic groups. Men performed significantly better than women across all ethnic groups. The remaining variables were assessed on subjects classified as white, consisting of 112,572 men and 102,080 women (together they represent 84.7 percent of self-labeled participants).

Of the subjects who performed the MRT, sexual orientation groups consisted of the following: heterosexual men 90.2 percent, gay men 5.5 percent, and bisexual men 4.3 percent; heterosexual women 90.2 percent, lesbians 2.9 percent, and bisexual women 6.8 percent. On the MRT, heterosexual men outperformed bisexual men, who in turn outperformed gay men. For women, the order of descending MRT performance was (from highest to lowest), lesbians, followed by bisexual women, followed by heterosexual women. No significant difference in the MRT scores was found between lesbians and bisexual women. These findings on the spatial abilities of lesbians relative to heterosexual women are supported by similar finds by van Anders & Hampson (2005) and contrary to earlier findings noted previously.

Height varied with respect to sexual orientation, with heterosexual men being taller than gay men, and gay men and bisexual men being of equal height. Lesbians were taller than heterosexual women, but of comparable height to bisexual women. In both sexes, the tallest group had the highest MRT average scores.

Women were found to have larger left 2D:4D ratios than men. Regardless of sex, subjects with smaller digit ratios scored higher on MRT than subjects with larger digit ratios.

The authors conclude that the consistency of the sex differences across the examined variables, together with evidence on hormonal involvement in MRT performance supports the view that the observed sex differences have a biological underpinning (Peters et al., 2007).

SUMMARY

No fewer than a dozen studies over the past decade have assessed the performance of gay men and lesbians as compared to heterosexual men and women on various tests of cognitive function. One report found no difference between heterosexual and gay men and between lesbians and heterosexual women on tests of verbal fluency and mental rotation. The remaining studies generally agree that gay men perform better than heterosexual men and similar to heterosexual women on tests of verbal fluency. On the other hand, gay men score lower than heterosexual men and once again similar to heterosexual women on mental rotations. Although several studies observed a trend for lesbians to perform similar to heterosexual men on both verbal fluency and mental rotations, the differences did not reach significance. In general these studies characterize the performance of lesbians on verbal fluency and mental rotation to be comparable to heterosexual women. Two subsequent studies with considerable statistical power (that is, large numbers of participants) have found that lesbians tend to perform like heterosexual men relative to heterosexual women on visual-spatial and mental rotation tests. The studies presented support the conclusion that the brains of gay men are less lateralized than the brains of heterosexual men, and the brains of lesbians are more lateralized than the brains of heterosexual women. Thus, overall these results are consistent with the neurohormonal theory, as some degree of prenatal androgen exposure appears to influence the neuronal circuitry that underlies cognition and sexual orientation.

Chapter 13

Gay Men and Older Brothers

THE FRATERNAL BIRTH ORDER EFFECT

A large number of studies over the past twelve years have shown that gay men have more older brothers than do heterosexual men (see Blanchard, 1997, 2001, and 2004 for reviews). Many of these studies were conducted using data collected in Canada, the Netherlands, England, and the United States. In contrast, older sisters do not influence the sexual orientation of their younger brothers. Furthermore, neither older brothers nor older sisters affect the sexual orientation of younger sisters; that is, lesbians do not have more older brothers and/or older sisters. This effect has been called the "fraternal birth order effect" (Blanchard, 1997). Results of a combined statistical analysis of 14 studies representing 10,143 male subjects convincingly show that gay men do, on average, have higher birth orders; that is, they are born later to the same mother than heterosexual men (Blanchard, 2004). This finding implies that the number of older siblings, or some factor related to this, must affect a newborn boy's sexual orientation.

Findings in support of this conclusion are as follows, and are noted in Blanchard (2001). Older brothers increase the probability of homosexuality in later born males. Available data indicate that each additional older brother increases the odds of homosexuality in a later-born male by 33 percent. Older sisters have no effect on the odds of homosexuality in later born males (but see Bogaert [1998a] for an exception). Furthermore, the interval between births does not influence the fraternal birth order effect. Neither older sisters nor older brothers affect the sexual orientation of later-born girls. Thus, girls appear to be invisible to the birth order effect. That is, girls do not affect the

Nature's Choice

sexual orientation of their siblings, and their siblings do not affect their sexual orientation.

Results from a study by Blanchard et al (2002) showed that fraternal birth order does not affect the age of the preferred erotic object. That is, the fraternal birth order effect is not seen in hebephiles, men who are sexually attracted to pubescent children (children who have just reached puberty), and pedophiles, men who are sexually attracted to prepubescent children.

A number of quantitative questions including the following have been addressed (Cantor, Blanchard, Paterson, & Bogaert, 2002). First, how strong is the fraternal birth order effect? Or, of all gay men, how many acquired their sexual orientation because of their birth order? Using epidemiological statistics, and the values of 2 percent to 3 percent for male homosexuality in the general population, it was determined that 14.8 percent to 15.2 percent of gay men can attribute their homosexuality to the fraternal birth order effect or, about 1 in 7. In addition, for men with 2.5 older brothers, the fraternal birth order effect accounts for half of the causative factors for a gay man's homosexuality. These finding are supported by a report on the population attributable fraction (PAF) defined as the proportion of gay men who can attribute their sexual orientation to their fraternal birth order (Blanchard & Bogaert, 2004). Combining two national probability samples, the PAF was calculated for 2,256 heterosexual men and 71 gay men and found to be 28.6 percent, but within the statistical limits of the 15.2 percent previously determined (Cantor et al., 2002). Last, the diversity of male populations in which the fraternal birth order effect has been observed makes it the most widespread factor in male homosexual development that has been identified (Cantor et al., 2002).

Birth Weight

Now comes the question of whether the birth order effect operates during gestation, early childhood, puberty, adolescence, or adulthood. The study that addresses this question is that of the birth order effect and birth weight. Blanchard & Ellis (2001) studied 3,229 men and women who weighed at least 2,500 grams (the minimum healthy birth weight equal to 5.5 pounds) at birth and whose mothers knew the sex of the child (or fetus) for each pregnancy prior to the study subject. Data were collected by questionnaires completed by the study subject's mother. The results revealed that boys with older brothers weigh less at birth than boys with older sisters. The average birth weights were not different for either lesbians versus heterosexual women or gay versus heterosexual men without older brothers or with older sisters only. Nonetheless, gay men with older brothers weighed about 170 grams (equal to 0.37 pounds or 6 ounces) less at birth than did heterosexual men with older brothers. In contrast, gay men and heterosexual men with no older brothers, or older sisters only, did not differ in birth weight. The correlation between the fraternal birth order effect and low birth weight was confirmed in a subsequent study (Blanchard et al., 2002). Collectively, these findings suggest that prior male pregnancies influence the development of subsequent male fetuses, and that those fetuses most strongly affected have lower birth weights and are more likely to be homosexual as adults. Thus, the birth order effect operates prior to an individual's birth; that is, the sexual orientation of later born males is determined prior to birth.

MATERNAL IMMUNE HYPOTHESIS

The favored hypothesized mechanism of the birth order effect is the "maternal immune hypothesis," first put forward by Blanchard & Bogaert (1996b). It states that the birth order effect may reflect a maternal immune response to factors associated with a male fetus, which becomes stronger with each additional pregnancy with a male fetus. The hypothesis rests on the immune system being capable of increasing its response to subsequent exposures to what it sees as foreign antigens; that is, it can "remember" that it has seen a specific antigen before. Each subsequent pregnancy with a male fetus is like a booster shot for a vaccination. Furthermore, the antigen that the mother's immune system remembers is associated with male fetuses. It has been suggested that the male fetal antigen is likely associated with proteins coded for by genes on the Y chromosome and, furthermore, may be the male specific, Y-linked minor histocompatibility antigen, referred to as H-Y antigen. Male fetuses carry the H-Y antigen from a very early stage of fetal development, but female fetuses do not. So, a mother's immune system could selectively respond to the H-Y antigen and produce antibodies to it. These antibodies could cross the placental barrier and access the fetal brain, and there alter the development of the brain and the subsequent sexual orientation of the individual. Since the immune response of the mother increases with each successive male pregnancy, the probability of homosexuality increases with each older brother.

Support for the maternal immune hypothesis comes from animal studies that have shown that the maternal immune system does respond to the fetal H-Y antigen, and that H-Y antibodies are found in the sera of mothers and male fetuses. In addition, male fetuses are also more likely than female fetuses to induce a maternal immune response to the rhesus (Rh) blood factor. Furthermore, animal studies demonstrate that 90 percent of male offspring of female mice actively immunized against the fetal H-Y antigen show reduced male-typical reproductive behavior (Singh & Verma, 1987).

How likely is it that brain tissue would be a target of H-Y antibodies? It actually turns out that the H-Y antigen is strongly expressed on the surface of brain cells (Blanchard & Bogaert, 1996b). Could anti-H-Y antibodies affect the sexual differentiation of the brain without affecting the differentiation of other tissues, specifically the genitalia? Studies of mice reveal that the testes can develop in the complete absence of H-Y antigen. It is therefore plausible that the genitalia could develop normally in human males whose H-Y antigen is only partially bound by anti-H-Y antibodies.

Can the maternal immune response hypothesis explain the low birth weight of gay men with a number of older brothers? Animal studies that have shown that maternal immunization with paternal, (male) antigens can affect fetal weight, placental weight, or both. Thus, it is possible that antimale antibodies produced in response to the progressive immunization of mothers by male antigens could decrease the birth weight of later born males in addition to increasing their probability of being homosexual.

The maternal immune hypothesis also implies that gay men with no or few older brothers acquired their sexual orientation from other causes, for example, fetal genes. As noted by Blanchard et al. (2002) it is also possible that maternal genes could play a role in the maternal immune response and thus influence the sexual orientation of a later-born boy.

OLDER BROTHERS AND OTHER CHARACTERISTICS

Handedness

Handedness (whether one is right-handed or left-handed) is sexually dimorphic, with more men being left-handed (also non-right-handed) than women. As we have seen (Chapter 10), gay men and lesbians tend to be more non-right-handed than heterosexual men and women. In a large meta-analysis the odds of non-right-handedness were 34 percent higher for gay men than for heterosexual men (Lalumière et al., 2000). Subsequent studies of handedness in gay men have been somewhat inconsistent, but generally support the higher rates of non-right-handedness in gay men.

A large meta-analysis was undertaken to determine if there is an interaction between the fraternal birth order effect and non-right-handedness in gay men (Blanchard, Cantor, Bogaert, Breedlove, & Ellis, 2006). Data from five studies were combined, generating all-male subjects comprised of 1,774 right-handed heterosexuals, 287 non-right-handed heterosexuals, 928 right-handed gay men, and 157 non-right-handed gay men. Data on each subject was generally obtained through self-administered questionnaires. Analysis of the data showed a significant interaction of handedness and older brothers, such that (1) the positive correlation between homosexuality and greater numbers of older brothers held only for right-handed males; (2) among men with no older brothers, homosexuals are more likely to be non-right-handed than heterosexuals, and among men with one or more older brothers, homosexuals are less likely to be non-right-handed than heterosexuals; and (3) the odds of homosexuality are higher for men who are non-right-handed or who have older brothers, relative to men with neither of these features, but odds for men with both features are similar to the odds for men with neither.

The authors suggest that these findings have at least two possible explanations: (1) the causative factors associated with non-right-handedness and older brothers (hypothesized to be hyperandrogenization and antimale antibodies respectively) counteract each other, yielding the functional equivalent of typical masculinization, or (2) the number of non-right-handed homosexuals with older brothers is smaller than expected because the combination of older brothers with the non-right-handedness factor(s) is toxic enough to lower the probability that the affected fetus will survive.

A further study tested the prediction that non-right-handed gay men will report fewer than expected older brothers (Blanchard & Lippa, 2006b). The participants were 2,486 heterosexual and homosexual, right-handed and non-right-handed male and female adults. Data on sibling composition, sexual orientations, and hand preference were collected from self-administered questionnaires. The sibling sex ratio defined as the ratio of older brothers to older sisters was calculated for each participant. The ratio of male live births to female live births in human populations is 106:100. Therefore, the ratio of older brothers to older sisters should approach 106 (older brothers per 100 older sisters). The sibling sex ratio for non-right-handed gay men was 83, which is significantly lower than the human sex ratio of 106. Conversely, the sibling sex ratio for right-handed gay men was 125, which is significantly higher than the expected ratio of 106. Thus, it appears that the effect of older brothers on sexual orientation is limited to right-handed men. The authors suggest two possible explanations. First, that older brothers increase the odds of homosexuality in right-handed males, but decrease the odds of homosexuality in non-right-handed males. Or second, that older brothers decrease the probability that non-right-handed gay men will be represented in the survey. This could be the result of the combination of some biological factor associated with older brothers and some biological factor associated with non-right-handedness being so toxic that it kills the fetus, or precludes his participation in surveys.

An additional study asked whether the different results observed for non-right-handed men have to do with heterosexual non-right-handers or the homosexual non-right-handers (Blanchard, 2007). Using a combined participant pool of 8,201 individuals, the sibling sex ratio was once again determined. The ratio for heterosexual right-handers was 105, for gay right-handers 128, for heterosexual non-right-handers 127, and for gay non-right-handers 96. The values for the gay right-handers and heterosexual non-right-handers were statistically different from the expected value of 106. These data support the conclusion that both heterosexual and gay non-right-handers contribute to the older brothers, handedness, and sexual orientation interaction.

A large Internet-based study has confirmed the previous results (Blanchard & Lippa, 2006a). Participants were 87,798 men and 71,981 women who completed questionnaires in a Web-based research project. Relationships between sexual orientation, fraternal birth order, and hand preference were explored. Results confirmed prior findings that (1) non-right-handedness is associated with homosexuality in both men and women, (2) older brothers increase the odds of homosexuality in men, and (3) the effect of older brothers on sexual orientation is limited to right-handed men.

Cognitive Functions

Gay men have been shown to display cognitive functions that are more similar to heterosexual women than to heterosexual men. Furthermore, they tend to display a feminized psychological gender. Thus, we can ask, does the fraternal birth order effect account for behavioral markers known to be strongly associated with male sexual orientation (Rahman, 2005)? If this is true, then the number of older brothers should strongly correlate with spatial ability and psychological gender among gay men as compared to heterosexual men. Participants were 214 individuals consisting of 89 heterosexual men, 80 gay men, and 54 heterosexual women. Participants completed questionnaires to assess sexual orientation, sibling sex ratio, and psychological gender and mental rotation tests. A significant fraternal birth order effect was observed, along with the expected differences in performance on mental rotation tests; heterosexual men scored significantly better than heterosexual women and gay men. No significant difference was found between gay men and heterosexual women. A significant difference was also found in psychological gender; heterosexual men scored higher—that is, they were more masculine— than gay men and heterosexual women. Furthermore, a significant positive correlation was found between mental rotation and psychological gender scores; as masculinity increased, so did mental rotation scores. These data confirm the strong cross-sex shift (female-typical) shown by gay males in mental rotation and psychological gender relative to heterosexual men. However, no correlation was observed between the number of older brothers a gay man had and mental rotation performance or psychological gender. That is, the fraternal birth order effect does not appear to be related to the typical behavioral feminization observed in gay men. The author suggests that the data could indicate that the fraternal birth order effect may act through an alternative neurodevelopmental pathway; the fraternal birth order effect may act independently from the prenatal androgen exposure pathway.

Environmental Effects

One issue that comes up with regard to the fraternal birth order effect is whether the underlying causative events occur during gestation or are a consequence of social factors related to being reared with older brothers. This question was examined in a study of prenatal versus postnatal (social/rearing) influences (Bogaert, 2006). A total of 944 individuals participated, of which 905 were not twins. The distribution of the 905 men with respect to sexual orientation was 329 heterosexual, 151 bisexual, and 425 homosexual. Approximately 45 percent of the participants were adopted, and the remainder was raised in some form of nonbiological or blended family. Data on birth order, sibling composition, and conditions of rearing were collected and analyzed. Overall, only biological older brothers (reared together or not) and no other sibling characteristic, including nonbiological older brothers and the time reared with older biological brothers or nonbiological brothers, predicted men's sexual orientation. The author concluded that these results support a prenatal origin to sexual orientation development in men and indicate that the fraternal birth order effect is probably the result of a maternal "memory" for male gestations or births.

Sex-Typical Behavior/Gender Role

Three additional studies have extended the characterization of the fraternal birth order effect. Bogaert (2003c) addressed the question of whether the birth order effect predicts the psychological attraction and sexual behavior in gay men. The study replicated the finding that the fraternal birth order effect predicts same-sex attraction in men, with each additional older brother increasing the probability of homosexual attraction by an average of 33 percent. The fraternal birth order effect was also independent of sexual behavior, including early same-sex behavior. Last, no sibling characteristics predicted sexual orientation in women. The author concluded the results suggest that experience-based theories (i.e., early same-sex play) of the fraternal birth order effect are not likely to be correct.

A second study by Bogaert (2003b) examined the relationship between the fraternal birth order effect and childhood sex-typing in a sample of 1,000 gay and heterosexual men derived from Kinsey Institute database. No relationship was found between fraternal birth order and childhood sex-typing (i.e., sex-typical behavior in childhood). The author concluded that postnatal (learning) and family/social environment mechanisms are unlikely to underlie the older brother effect in men. Furthermore, a study of gender role/identity and the fraternal birth order effect found no relationship (Bogaert, 2005).

Height and Weight

The relation between fraternal birth order effect and body size has also been examined (Bogaert, 2003a). A correlation between fraternal birth order and height was observed, with a homosexual orientation most likely to occur in men with a high number of older brothers and short stature. No significant effects were observed for weight. The author concluded that the fraternal birth order effect impacts physical development, such that it lasts and is detectable into adulthood.

The question of whether parental age has an affect on the fraternal birth order effect was addressed in a study of 4,690 participants: 2,189 heterosexual men, 70 gay men, 2,380 heterosexual women, and 51 lesbians (Bogaert & Cairney, 2004). The authors concluded that a positive association existed between the number of older brothers and the likelihood of male homosexuality, but this association weakened with increasing parental age. In addition, this interaction was largely carried by the relationship of the mother's age to the older sibling effect. That is, the fraternal birth order effect is stronger in men with older brothers born to younger mothers.

An additional study of 302 gay men and 302 heterosexual men matched for year of birth—year of birth has been found to be a confounding variable in birth order research (Hare & Price, 1969)— found a relationship between the fraternal birth order and height (Bogaert & Liu, 2006). Gay men's lesser height relative to heterosexual men's is more pronounced with a high number of older brothers. Additional statistical analysis revealed an independent relationship between fraternal birth order and parental age. Having younger parents increases the likelihood of a homosexual orientation in men. The authors suggest that the fraternal birth order effect may reflect a biological mechanism that affects not only sexual orientation but also aspects of physical growth and development. Why there is an independent interaction between the fraternal birth order effect and parental age is unknown.

SUMMARY

The fraternal birth order effect is the finding that older brothers increase the odds of homosexuality in later-born males by as much as 33 percent. Older sisters and younger siblings have no effect. No birth order effect has been observed in women. Approximately 1 in 7 to 1 in 4 (i.e., 17 percent to 25 percent) gay men can attribute his sexual orientation to the fraternal birth order effect. A study of birth weight in gay men with older brothers confirms that they weigh less at birth than do heterosexual men with older brothers. Thus, it was concluded that prior male pregnancies influence the development of subsequent fetuses, and that those fetuses most strongly affected have lower birth weights and are more likely to be homosexual as adults.

This last conclusion has led to the maternal immune hypothesis, which states that the birth order effect may reflect a maternal immune response to factors associated with a male fetus, which becomes stronger with each additional pregnancy with a male fetus. At present, there are not enough data to confirm the maternal immune hypothesis as a mechanism for the fraternal birth order effect. However, the data do suggest that it occurs during gestation. Furthermore, since birth weight is clearly related to prenatal events and is predicted by the interaction between the number of older brothers and sexual orientation, it follows that both the fraternal birth order effect and sexual orientation are prenatally determined.

An interaction between handedness and the fraternal birth order effect has been characterized as follows:

1. There is a positive correlation between homosexuality and greater numbers of older brothers, but only for right-handed males.
2. Among men with no older brothers, gay men are more likely to be non-right-handed than heterosexual men; among men with one or more older brothers, gay men are less likely to be non-right-handed than heterosexuals.
3. The odds of homosexuality are higher for men who have a non-right-handed preference or who have older brothers, relative to men with neither of these features, but odds for men with both features are similar to the odds for men with neither.

These observations are thought to have at least two possible explanations:

1. The causative factors associated with non-right-handedness and older brothers counteract each other, leading to typical masculinization.
2. The number of non-right-handed gay men with older brothers is smaller than expected because the combination of older brothers with the non-right-handedness is toxic enough to lower the probability that the fetus will survive.

Differences in cognitive functions for gay men with older brothers have been examined. No relationship was observed between the fraternal birth order effect and the cognitive function of mental rotation. Furthermore, no relationship was observed between psychological gender and the fraternal birth order effect. These data suggest that the prenatal androgen exposure pathway that is thought to lead to feminization of cognitive function and psychological gender in gay men may operate independently from the fraternal birth order effect.

Data on birth order, sibling composition, and conditions of rearing indicate that only biological brothers (reared with or not) and no other sibling characteristic, including nonbiological brothers and time reared with older brothers or nonbiological brothers, predicted men's sexual orientation. Thus, family and social environment mechanisms are unlikely to underlie the fraternal birth order effect.

There is a relationship between the fraternal birth order and height such that gay men with older brothers are shorter than heterosexual men. Having a younger mother increases the likelihood of homosexuality in men.

Last, the diversity of male populations in which the fraternal birth order effect has been observed makes it the most widespread factor in male homosexuality development that has been identified.

Chapter 14

Maternal Behaviors

A recent study by Ellis and Cole-Harding (2001) examined the independent and combined effects of maternal stress, alcohol consumption, and cigarette smoking on the sexual orientation of offspring. For the sake of simplicity, each will be examined separately, and finally their combined effects discussed. Last, the effects of gonadal steroids and drugs that interact with their receptor systems will be discussed.

STRESS

Six studies of the prenatal effects of maternal stress on the sexual orientation of offspring have been reported. The first five of these concerned the sexual orientation of male offspring. The first two studies appeared in the early 1980s and reported higher than normal rates of male homosexuality/bisexuality for mothers who recalled experiencing severe stress during pregnancy (Dorner et al., 1980; Dorner, Schenk, Schmiedel, & Ahrens, 1983). The effects were small, but significant, and were criticized on methodological grounds. A third study found a significant correlation between the frequency of stressful events plus their subjective severity during the second trimester and homosexuality in male offspring (Ellis, Ames, Peckham, & Burke, 1988). The authors suggest that the impact of prenatal stress appears to depend not only on the intensity and duration of the stress, but also on its controllability and predictability. This is supported by observations in rats, for which the release of stress hormones is greater when the stress is beyond the animal's ability to control or anticipate it. Two additional studies failed to replicate the three studies

Nature's Choice

that found a positive correlation between maternal stress and male offspring homosexuality (Schmidt & Clement, 1990; Bailey, Willerman, & Parks, 1991). The major criticisms of the first five studies are that they were based on small sample sizes, were not representative of the general population, and the experimental designs were retrospective.

The most recent study by Ellis and Cole-Harding (2001) was conducted over a 10-year period, and based on questionnaires from mothers and their college-age offspring, which included 2,554 males and 5,051 females. Sexual orientation of the participants was assessed by questionnaire. The questionnaire was designed to assess consistency of answers, and those found to provide inconsistent answers were not included in the study. Prenatal stress was also assessed in detail through a 10-page questionnaire. Seven categories of 76 potentially stressful events were assessed over 10 3-month blocks of time starting 18 months prior to pregnancy through 3 months following birth. The severity of stressful experiences was rated by the mothers on a scale of 1 (minimal) to 9 (most severe stress imaginable). The reliability of the mother's reports of stress during her pregnancy was assessed by the completion of a second questionnaire a minimum of two months after the first was completed. Agreement between the two responses was highly significant statistically.

The average monthly levels of prenatal stress reported by mothers of gay men were greater than that experienced by the mothers of heterosexual men. A statistically significant correlation was observed between maternal stress during the second month of pregnancy and homosexuality in male offspring. The maternal stress effect on male offspring homosexuality is considered a small effect. In contrast, no significant correlation between maternal stress and homosexuality in female offspring was observed.

It has been reported that male embryos (XY chromosomes) develop faster than do female embryos (XX chromosomes) at very early stages of development (Mittwoch, 2000). The authors suggest that maternal stress may reduce the rate of fetal development in ways that slow down brain masculinization/defeminization.

Because the maternal stress effect is small and homosexuality is comparatively rare, the authors note that improving on their experimental design would require exceedingly large samples, at least

25,000 mothers and their offspring, who would have to be located 20 years after birth in order to determine their sexual orientation.

ALCOHOL

Using the same population of mothers and their offspring, Ellis and Cole-Harding (2001) examined the effects of alcohol consumption on the sexual orientation of their offspring. They noted that alcohol readily crosses the placenta from the mother's blood to the fetus' blood supply. Alcohol can also affect other systems in ways that resemble stress. In other studies, the male offspring of female rats that consumed alcohol during pregnancy were more likely to exhibit feminine **receptive** postures at puberty than were control males. This effect was observed independent of and in combination with prenatal stress.

The anticipated result based on these observations was that prenatal alcohol consumption would impact male offspring such that their adult sexual behavior would be feminized/demasculinized. Furthermore, the greatest degree of feminization/demasculinization of male offspring would result from maternal stress plus alcohol consumption. It was further anticipated that the most vulnerable time for these effects would be early in the second trimester of pregnancy.

The same population and protocol for data collection and analysis was used as for the study of maternal stress. The analysis of the effects of prenatal alcohol consumption on the sexual orientation of offspring over each month of pregnancy showed no effects greater than chance for either male or female offspring. That is, no correlation was found between maternal alcohol consumption and homosexuality in either males or females.

NICOTINE

Ellis and Cole-Harding (2001) also examined the effects of smoking during pregnancy on the sexual orientation of both male and female offspring using the same population of subjects and their mothers as for the studies of stress and alcohol. Animal studies suggest that nicotine exposure would feminize/demasculinize male offspring's adult sexual behavior. Analysis of the data revealed significant differences between heterosexual and homosexual offspring for females, but not for males. For females, mothers of lesbians smoked significantly more throughout pregnancy than did mothers of heterosexual women. Rigorous statistical analysis of the data revealed that only the first two months could be considered statistically significant. It was suggested that the results might be a consequence of nicotine, a known stimulant, speeding up the developmental processes of a female fetus in ways that tend to masculinize/defeminize those brain circuits that support sexual orientation behavior. Furthermore, in support of the results, one study revealed that smoking was positively associated with the mother's circulating testosterone levels and the testosterone levels of female offspring (Kandel & Udry, 1999).

COMBINED EFFECTS OF STRESS, ALCOHOL, AND NICOTINE

Further statistical analysis of the data was conducted to assess the combined effects of stress, the mother's perceived ability to control prenatal stress, alcohol consumption, and nicotine use on the sexual orientation of both male and female offspring. The mother's perception of her ability to control prenatal stress was not significantly related to the sexual orientation of either males or females. Only prenatal stress during the second month of pregnancy was associated with an elevated probability of male offspring being homosexual or bisexual. For female offspring, cigarette smoking by their mothers during the first month of pregnancy was the best predictor of female homosexuality or bisexuality. Cigarette smoking during the first month of pregnancy was joined by prenatal stress during the fourth month as a significant covariant of sexual orientation, with values for both variables increasing the probability of homosexuality or bisexuality in female offspring.

HORMONES

The ovaries primarily produce two major classes of steroid hormones, estrogens, and progestins. They also produce low levels of testosterone. The primary estrogen produced is estradiol; others include estrone and estriol. Progestins, primarily progesterone, are produced in the corpus luteum of the ovary and are important for uterine, vaginal, and mammary gland growth. Both are important in establishing and maintaining pregnancies and both influence female behavior.

Diethylstilbestrol (DES)

Diethylstilbestrol (DES) is a nonsteroidal synthetic estrogen. From the 1940s through the 1960s it was given to women with problem pregnancies. Its use was halted when it was found to have adverse effects, including an increased risk of cancer. Its chemical structure allows it to access the fetal circulation; it does not masculinize the genitalia, but does reach the fetal brain in both rodents and nonhuman primates (Slikker Jr., Hill, & Young, 1982). Remember that the masculinizing effects of androgens in the brain are ultimately mediated by estradiol within target neurons. Studies of women exposed to DES during fetal development show that they are more likely to have homosexual or bisexual fantasies than women not exposed to DES (Meyer-Bahlburg et al., 1995). A comparable study of men exposed to DES during gestation found no influence of DES on sexual orientation (Meyer-Bahlberg & Ehrhardt, 1996).

Progestins

The picture for progestins is a bit more complicated, as they can either inhibit or enhance the effects of androgens, depending on the specific chemical structure of the progestin. This is a consequence of most progestins binding not only to progesterone receptors, but also to androgen and estrogen receptors. Animal studies have shown that gestational treatment with various progestins alters normal physical and behavioral development; that is, it can cause masculinization of females (Bardin, 1983) or demasculinization of males (Wright, Giacomini, Riahi, & Mowszowicz, 1983).

Thus, it is clear that *no* steroid or nonsteriodal drug that interacts with gonadal steroid-dependent systems should be administered to a woman during pregnancy.

PRESCRIPTION DRUGS

The effects of prescription drug use during pregnancy on the sexual orientation of offspring was assessed in a questionnaire study of 5,102 mothers, who provided data on their use of 19 drug categories and the sexual orientation of their offspring (Ellis & Hellberg, 2005). Comparisons were made between mothers of heterosexuals and mothers of homosexuals. Five drug categories prescribed to mothers during pregnancy were significantly related to the offspring's sexual orientation. The authors note that the sample sizes for most categories of drugs were exceedingly small, and thus caution must be exercised in offering interpretations.

Three drug types affected male offspring with respect to sexual orientation: antinausea and vomiting medication, gamma globulin, and amphetamine-based diet pills (primarily Dexedrine and Tenuate Dospan). Three drug types also affected female offspring with respect to sexual orientation: amphetamine-based diet pills, DES (diethylstilbestrol), and synthetic thyroid medications, primarily Synthroid and thyroxine. Upon controlling for maternal age, maternal education, and maternal recall, none of the classes of drugs affecting male sexual orientation remained statistically significant. For females, two classes of drugs remained statistically significant after controlling for maternal characteristics, prescription diet pills and synthetic thyroid medications. After controlling for maternal characteristics, the DES effect was no longer statistically significant.

The other parameter examined was the months during pregnancy when the effects of the drugs were most significant. Maternal consumption of drugs was comparable for mothers of male homosexuals, male heterosexuals, and female heterosexuals. For the mothers of female homosexuals, high rates of consumption of prescription drugs occurred during the second month of pregnancy. A similar trend in the same direction was observed during the third month as well. The authors conclude that female offspring are more vulnerable to alterations in sexual orientation via exposure to prescription drugs, and, further, this vulnerability is greatest during the first trimester. Since this is the first study of the effects of prescription drugs on the sexual orientation of offspring, these results beg for confirmation through additional studies.

SUMMARY

With regard to maternal behaviors, five areas have been examined: maternal stress, and consumption of alcohol, nicotine, steroidal hormones, and prescription drugs. Fairly good data support a small effect relationship between maternal stress, particularly in the second month of pregnancy, and homosexuality in male offspring.

No effect of prenatal alcohol consumption was found over the course of pregnancy on the sexual orientation of either male or female offspring. Nicotine consumption (smoking) did affect female offspring, but not male offspring. Mothers of lesbians smoked significantly more throughout pregnancy than did mothers of heterosexual women; however, only the first two months were statistically significant. Furthermore, smoking may be positively associated with a mother's circulating testosterone levels and the testosterone levels of female offspring.

Cigarette smoking during the first month of pregnancy and prenatal stress during the fourth month combined to increase the probability of homosexuality or bisexuality in female offspring.

Studies on the use of steroidal drugs that can mimic the actions of natural hormones, including estrogen, testosterone, progesterone and cortisol, are not conclusive with regard to their effects on offspring sexual orientation. However, the use of such drugs during pregnancy is not advisable.

Last, an initial study suggests that the consumption of certain classes of prescription drugs by pregnant women can impact the sexual orientation determination of their offspring. The drugs with greatest effect were prescription diet pills and synthetic thyroid medications. Furthermore, female offspring were more vulnerable to alterations in sexual orientation via exposure to prescription drugs and this vulnerability was greatest during the first trimester.

Chapter 15

Conclusions

Where does this leave us with regard to the biological origins of human sexual orientation? What can we conclude? Is sexual orientation biologically determined? Is it genetically determined and therefore inherited? What factors influence sexual orientation? Let's first review some general facts, and then review the science we have examined.

First and foremost, homosexuality is not a disease. Nor is it related to pedophilia, hebephilia, or childhood sexual abuse. Therefore it appears that it is a normal variant of human sexuality. Second, the fetus controls its own sexual development through its own gonadal and adrenal secretions of steroid hormones, most prominently testosterone and other androgens. In addition, the fetus is protected from circulating maternal hormones; under normal conditions, Mom hasn't done anything that can alter the fetus's sexual orientation. Finally, it seems reasonable at present to conclude that heterosexuality, bisexuality, and homosexuality are all biologically determined by normal developmental processes and comprise the spectrum of accessible human sexual orientations exclusive of asexuality, pedophilia, and hebephilia.

Animal experiments demonstrate that testosterone is a critical determinant of male sexuality and can influence female sexuality. Reduced availability of testosterone can invert male sexual behavior (males, with XY chromosomes display female sexual behavior) while increased availability of testosterone can invert female sexual behavior (females, with XX chromosomes display male sexual behavior). Similar inversions of sex-typical behaviors are also observed in female rodents and spotted hyenas that are exposed to high levels of testosterone during gestation under normal conditions. Furthermore, inversions of sexual orientation and sex-typical behaviors are observed in human conditions in which the function of testosterone is blocked in males (as in complete androgen insensitivity), or in which the availability of androgens is increased in females (as in congenital adrenal hyperplasia). All of these observations are consistent with the neurohormonal theory, which states that hormonal and neurological variables, operating during gestation, are the main determinants of sexual orientation.

What data do we have that support a biological origin of sexual orientation, and how well do the data fit the neurohormonal theory? First, let's see what the latest genetic and anatomic data contribute to our understanding of the biology of sexual orientation. Do the available genetic studies implicate genes in the determination of human sexual orientation? General population studies do demonstrate that sexual orientation runs in families. Heterosexual men have more heterosexual brothers than do gay men and vice versa; gay men have more gay brothers than do heterosexual men. The same is true of women and sexual orientation. Heterosexual women have more heterosexual sisters and lesbians have more lesbian sisters. If sexual orientation runs in families, it should be heritable. The strongest population genetic test of heritability is whether a trait has a significantly greater probability of being present in both monozygotic (identical) twins than in both dizygotic (nonidentical) twins. Twin studies of both gay men and lesbians demonstrate a near 50 percent probability of both monozygotic twins being homosexual, while the probability for dizygotic twins was less than half that value, 22 percent for men and 16 percent for women. Thus, homosexuality appears heritable.

If homosexuality is heritable, then genes are involved. At present we have four studies that have examined linkage of a gene in the Xq28 region of the X chromosome with a fraction of male homosexuality. Three studies, two from the same laboratory, support the linkage of Xq28 associated DNA sequences in 64 to 67 percent of gay brothers whose families displayed maternally inherited male homosexuality. A fourth study was unable to replicate these results, although there were some differences in the methods employed by the two groups. Linkage of male homosexuality to areas on chromosomes 7, 8, and 10 has also been demonstrated. So, does a gene or group of genes control sexual orientation? From all of the genetic data currently available, we can conclude that the answer is yes for both male and female homosexuality. Furthermore, it is likely that more than one gene is involved in sexual orientation determination. That is, sexual orientation is polygenetic. We can also say that there are likely genes that influence *heterosexuality* in both men and women. At present the best we can say about an X-linked gene influencing a fraction of male homosexuality is that it is controversial. Nonetheless, the genetic studies collectively support the neurohormonal theory.

Well, what about brain anatomy? Three anatomical studies report statistically significant differences in the average size of brain structures between heterosexual and gay men; two studies included heterosexual women for whom the brain areas were comparable to that of gay men and significantly different from heterosexual men. Two of these structures were within the hypothalamus, which animal studies have implicated in the regulation of sexual behavior. If sexuality is even partly biologically determined, we would expect that some area of the hypothalamus would show a difference in volume (in the total number of cells). The third area demonstrated to be different was the anterior commissure, which is very close to the hypothalamus and provides for visual, auditory, and olfactory communication between the left and right sides of the brain. In this case, the cross-sectional area was different between gay and heterosexual men and heterosexual men and heterosexual women. Replications of two of these studies were not completely successful, so there is still the need for confirmation of these findings. Nonetheless, these reports of correlations between brain structure size and sexual orientation taken together provide additional support for the neurohormonal theory.

Furthermore, the data on the anterior commissure support the expanded hypothesis of Allen and Gorski (1992) that factors operating early in development differentiate sexually dimorphic structures and functions of the brain in a global fashion that is throughout the brain. This hypothesis accounts for the determination of sexual orientation as well as sex-typical behaviors. This is a working hypothesis, not a proven fact. All of these studies have yet to be completely replicated, and many additional studies will be required before this hypothesis is fully tested and the conclusions can be considered fact. However, these studies are all that is available for the moment, and they support the idea that biology determines human sexual orientation. It should be noted that currently no studies present credible data refuting the claim that sexual orientation is biologically determined, nor has credible data been published identifying family or cultural environmental factors that influence sexual orientation.

Ultimately, an additional question will have to be addressed. Does the difference in brain structures observed between heterosexual and homosexual individuals cause the difference in sexual orientation, or does the difference in sexual orientation and the associated behaviors result in the observed differences in brain structure? To answer this question would require that we conduct a longitudinal study and measure these brain structures in a large number of individuals as a fetus, as a very young child, and as an adult. Then we could look to see if a correlation exists between the differences in the size of the areas as a function of the adult individual's sexual orientation. Such a study at present is not doable, and, if it were, would be complicated, requiring a large sample size, a large number of determinations, and a time span of more that 20 years, to say nothing of there not being, at present, any noninvasive methods for making such determinations. These factors would make it very expensive if it were technically possible to make such determinations on intact, live individuals, and, unfortunately, not likely to be funded.

A close relationship between sexual orientation and sex-typical behaviors has been established. Studies of both men and women reveal that gay men and lesbians both recall more cross-gender sex-typical behavior as children than do heterosexual men and women. Furthermore, a study of children revealed that a positive linear correlation exists between maternal circulating testosterone and masculine sex-typical behavior in girls. No relationship between maternal testosterone and sex-typical behavior was observed for boys. These observations are also consistent with the neurohormonal theory.

Anthropometrics, or the comparative study of human body measurements, has also contributed data on the biology of sexual orientation. Studies of the long bones of the hand and body demonstrate that they are sexually dimorphic, being longer in men than in women. Hand and body long bones were found to be longer in heterosexual men than in gay men. Moreover, they were also longer in lesbians than in heterosexual women. Studies of the 2nd to 4th finger length (2D:4D) ratio revealed that the ratio is sexually dimorphic; the ratio is near one for heterosexual women and less than one for heterosexual men. A relationship between 2D:4D ratio and prenatal androgens is supported by the finding that women and men with congenital-adrenal-hyperplasia had smaller, more masculine 2D:4D ratios than did women and men without congenital-adrenal-hyperplasia. Thus, prenatal exposure to androgens appears to result in lower 2D:4D ratios.

Six of eight studies of lesbians found that the 2D:4D ratio for lesbians was more similar to that of heterosexual men than that of heterosexual women and close to the ratio of women with congenital-adrenal-hyperplasia. Furthermore, in a study of identical twins, one of whom was lesbian and the other heterosexual, the lesbian twins had significantly smaller 2D:4D ratios than did their heterosexual twin. The lesbians in the study had 2D:4D ratios comparable to the population average for males. No significant differences in the 2D:4D ratio was observed when both twins were lesbian. Since the observed differences in the 2D:4D ratio for co-twins indicate that differences existed in their prenatal environment during the first trimester of development, and, further, that these differences were associated with sexual orientation, prenatal factors should be considered causative factors in sexual orientation development. These findings are consistent with the neurohormonal theory.

The picture for gay men is less clear, as two studies have found that the 2D:4D ratio of gay men is smaller than heterosexual men, four studies found the ratio to the same for both, and three studies found the ratio for gay men to be larger than that for heterosexual men. This finding contributes to the emerging idea that gay men can be hypomasculinized with respect to some traits and hypermasculinized with respect to others. We will have to await additional and likely more refined studies of gay men to provide a clearer understanding of the relationship of 2D:4D ratios and sexual orientation in men. The emergence of the fraternal birth order effect also complicates these kinds of determinations. Researchers will have to select their male subjects with care to differentiate between gay men with and without older brothers. The story for the 2D:4D ratio is also complicated by the differences not being consistent across different races.

Another anthropometric characteristic that is sexually dimorphic is the number of ridges and their pattern associated with fingerprints. Dermal ridges are influenced by genetics and subject to intrauterine environment during their development. The total number of ridges on both hands is higher in men than in women, and this is not due to the larger hand size of men. Some studies suggest that testosterone can influence total ridge counts. The consensus of the studies of total ridge counts in gay men is that there is no difference in total ridge count between gay men and heterosexual men. Two studies of ridge counts in lesbians have been conducted. One found no difference between the total ridge counts in lesbians versus heterosexual women. A second study examined total ridge counts in identical twins, one of which was lesbian and the other heterosexual. Comparing total ridge counts between each twin pair, lesbian twins were found to have lower total ridge counts than their heterosexual twins. Thus, at present, there is no consensus with regard to total ridge counts and sexual orientation in either men or women. Nonetheless, the positive finds are consistent with the neurohormonal theory.

Height and weight are also sexually dimorphic, with men being taller and heavier than women. Based on the neurohormonal theory, we might expect gay men to be feminized relative to heterosexual men and therefore be shorter and lighter than heterosexual men. Lesbians on the other hand might be expected to be masculinized and therefore be taller and heavier than heterosexual women. Five studies of height in men have been published; two studies found gay men to be shorter than heterosexual men, while three studies found no height difference between gay men and heterosexual men. Seven studies of weight in men have been published; four studies found gay men to be lighter than heterosexual men, while three studies found no difference in weight between gay men and heterosexual men.

Two studies of height in women have been published. One found that lesbians were taller than heterosexual women, while the other found no difference in height between lesbians and heterosexual women. Four studies have examined weight in women. Three reported lesbians to be heavier than heterosexual women, and one found no difference in weight between lesbians and heterosexual women. Thus, at present, the data in hand do not support the idea that gay men are shorter and lighter than heterosexual men. There is some support for the conclusion that lesbians are heavier than heterosexual women, but no support for the idea that lesbians are taller than heterosexual women.

Another anthropometric trait that is sexually dimorphic is waist-to-hip ratio (WHR). Men on average have larger waists and narrower hips than do women. Once again, if lesbians are masculinized, we might expect them to have a WHR more similar to heterosexual men than to heterosexual women. One study examined the WHR in lesbians that self-identified as either "butch," relatively masculine, or "femme," relatively feminine. Butch lesbians were found to have a higher WHR (a more masculine build) than either femme lesbians or heterosexual women. Furthermore, the WHR of lesbians correlated with their recalled childhood sex-typical behavior (tomboyness) and their adult levels of testosterone. These data are consistent with predictions for lesbians based on the neurohormonal theory.

Another anthropometric measure that has been studied is penis size. One might expect that if gay men are more feminized relative to heterosexual men, that they would have smaller penises. Two studies agree, contrary to prediction, that gay men in fact have somewhat larger penises than do heterosexual men. As noted before, an emerging picture of gay men is that relative to heterosexual men they can be hypomasculinized on some measures and hypermasculinized on other measures. Penis size would appear to be one measure on which gay men are hypermasculinized relative to heterosexual men.

Onset of puberty is also different for males versus females. In general, boys reach puberty later than girls. Once again, based on the neurohormonal theory we might expect gay men to reach puberty earlier than heterosexual men and lesbians to reach puberty later than heterosexual women. Studies have confirmed that gay men reach puberty earlier than do heterosexual men. In contrast, however, no difference in the age of onset of puberty has been observed for lesbians relative to heterosexual women. Thus, the data for men are consistent with the neurohormonal theory prediction, but the data for women are not.

Another function that is sexually dimorphic in humans is hearing and related response of the ear. The ear is the sensory organ that allows us to hear, but it also emits sounds spontaneously. Two types of these sexually dimorphic spontaneously emitted sounds are spontaneous otoacoustic emissions, or SOAEs, and click-evoked otoacoustic emissions, or CEOAEs. SOAEs are more prevalent in females than males, while CEOAEs are stronger in females than males. Hearing sensitivity is also better in females than males and is better in right than in left ears. Both otoacoustic emissions are strongly influenced by genetics, and they have high heritability. Studies of nonidentical twins suggest that androgens masculinize the OAEs of females with a male twin. A study of OAEs in both heterosexual and homosexual men and women revealed that the OAEs of lesbians and bisexual women were shifted toward the male range and masculinized relative to heterosexual women. No difference was found between OAEs of gay men versus heterosexual men.

A second way of monitoring hearing functions is to record auditory evoked potentials (AEPs) from the scalp of individuals. Sex and ear differences exist in certain AEPs in both infants and adults. Lesbian's AEPs were masculinized relative to heterosexual women, consistent with the neurohormonal theory. Gay men on the other hand displayed hypermasculinized AEPs relative to heterosexual men, lending additional support to the idea that the underlying mechanisms of sexual orientation determination are different for men and women.

A third sensory response that is sexually dimorphic is the eye-blink startle reflex, which is assessed through a response known as prepulse inhibition (PPI). Women display a lower PPI than men. A study of heterosexual and homosexual men and women revealed that lesbians displayed a level of PPI close to that for heterosexual men; their responses appeared masculinized relative to heterosexual women. The PPIs of gay men on the other hand were not significantly different than that of heterosexual men.

The responses of lesbians, on all four measures, SOAEs, CEOAEs, AEPs, and PPI were masculinized relative to heterosexual women. Since evidence suggests that these responses develop during gestation and appear dependent on androgens, these observations are consistent with the neurohormonal theory. In contrast, the responses of gay men were mixed, with AEPs hyper-masculinized relative to heterosexual men. No significant differences in OAEs and PPIs were observed between gay men and heterosexual men. These findings suggest that the effects of androgens on the sexual differentiation of gay men are not simply linear, and identify another measure on which gay men are hypermasculinized.

A recently identified measure on which men and women differ is their response to smelling specific steroidlike compounds that are found in sweat and urine. The response of men and women to smelling these compounds is to selectively activate different areas of the hypothalamus. Women activate the preoptic ventromedial hypothalamus when smelling the male sweat compound AND, while men activate the anterior hypothalamus when smelling the female compound

EST. These areas are both involved in sexual behavior. Thus, AND and EST appear to be human pheromones. Studies of gay men and lesbians found that gay men respond to smelling AND, similar to heterosexual women, by activating the preoptic ventromedial hypothalamus, while lesbians respond similar to heterosexual men by activating the anterior hypothalamus when smelling EST. Once again gay men and lesbians display a cross-sex shift in their response to proposed human pheromones. Activation of these areas of the brain has also been shown to lead to physiological responses related to heightened physical arousal.

Another functional trait on which men differ from women is cerebral lateralization, or the degree to which selected motor and cognitive functions are associated with one hemisphere or the other. In general, men are more lateralized, that is, functions are more restricted to one hemisphere. In women, selected functions are more distributed across both hemispheres.

Handedness, whether one is right-handed or left-handed, is a measure of left or right lateralization of motor functions. Left-handedness in the general population is about 10 percent, with slightly more men than women being left-handed. Handedness is likely determined well before birth. Thus, once again, one might expect more homosexual individuals to be left-handed. Studies have revealed that more lesbians are non-right-handed than are heterosexual women, consistent with the neurohormonal theory. In contrast, studies of gay men have not yielded consistent results. Thus, no definitive conclusions are warranted on the relationship between handedness and sexual orientation in men. As we have seen the characteristics of men whose homosexuality can be attributed to the fraternal birth order effect may confound the issue of handedness in gay men. This point awaits clarification.

The other aspect of cerebral lateralization that has been studied is cognitive functions. The left hemisphere excels at sequential, intellectual, analytical, rational, verbal thinking and motor functions. The right hemisphere excels at sensory discrimination and visual-spatial, emotional, intuitive, abstract and nonverbal thinking. Data from a number of studies in animals, human dizygotic twins, and humans with endocrine disorders support the hypothesis that exposure to androgens during gestation contributes to the lateralization differences observed between men and women. Women perform better than men on tests of verbal fluency, while men perform better than women on tests of mental rotation. Gay men have been found to perform more similar to heterosexual women than heterosexual men on tests of both verbal ability and mental rotation. Although a trend for lesbians to perform similar to heterosexual men on tests of verbal fluency and mental rotation has been observed, only recently have studies found the differences to be statistically significant. Thus, the performance of lesbians on tests of verbal fluency and mental rotation appear to be more similar to heterosexual men than to heterosexual women.

Another important factor that has been shown to influence sexual orientation in men is the number of older brothers a man has. This has been called the fraternal birth order effect. Older brothers increase the probability of homosexuality in later born males by 33 percent. Older sisters have no effect on the sexual orientation of either later born males or females. Statistical analyses indicated that one in seven to one in four gay men can attribute their homosexuality to the fraternal birth order effect. Furthermore, a correlation exists between homosexuality in men with older brothers and lower birth weight. This suggests that the fraternal birth order effect operates during gestation. At present the hypothesized mechanism by which the fraternal birth order effect influences the sexual orientation of men with older brothers is the maternal immune hypothesis. This hypothesis states that each successive male pregnancy causes the mother's immune system to make antibodies against male antigens, and these antibodies influence birth weight and sexual orientation development. A number of studies suggest that the fraternal birth order effect and prenatal androgen exposure may be two separate paths for the development of homosexuality in men. Last, the diversity of male populations in which the fraternal birth order effect has been observed makes it the most widespread factor in male homosexual development identified thus far.

Another area of investigation into the causes of homosexuality is maternal behaviors or how maternal stress and drug use during pregnancy impact the sexual orientation of their offspring. A recent large study of maternal stress and the sexual orientation of both male and female offspring revealed that maternal stress in the second month of gestation contributed significantly to homosexuality in male offspring. No significant relationship between maternal stress and homosexuality in female offspring was observed.

The effect of alcohol consumption during pregnancy on the sexual orientation of offspring showed no effects greater than chance for either male or female offspring. A similar study of the effects of maternal cigarette smoking on the sexual orientation of offspring revealed that cigarette smoking during the first two months of pregnancy increased the incidence of homosexuality in female offspring, but not in male offspring. This effect is thought to be the results of the fact that nicotine, the primary active ingredient in cigarette smoke, speeds up fetal development, and that cigarette smoking is positively associated with the mother's circulating testosterone levels and the testosterone levels of female offspring. Further examination of the combined effects of maternal stress, alcohol, and nicotine revealed that only maternal stress during the second month of pregnancy was associated with an elevated probability of homosexuality in male offspring. For female offspring, maternal cigarette smoking during the first month of pregnancy was the best predictor of female homosexuality or bisexuality.

Diethylstilbestrol (DES) is a nonsteroidal synthetic estrogen that can access fetal circulation when administered to pregnant women. At present, there are conflicting results and therefore no consensus on the effects of DES on the development of sexual orientation.

SUMMARY OF POSITIVE AND NULL EFFECTS

Perhaps the easiest way to generate a picture of just how much data we have on the biological origins of sexual orientation is to tabulate the identified measures associated with homosexuality in men and women (Table 15.1).

On all of these measures, except genetics, penis size of course, and the birth order effect, lesbians are masculinized, and masculinization on these measures is associated with prenatal exposure to testosterone. Thus, homosexuality in women appears to be consistently associated with increased testosterone as predicted by the neurohormonal theory.

On five of the measures, gay men are feminized, while on two of them, gay men are hypermasculinized. Thus, it appears that the mechanism(s) that lead to homosexuality are more varied in men than in women. This has led to the idea that in men there is a relatively narrow range of testosterone levels that provide for heterosexual male

TABLE 15.1. Biological Measures Associated with Homosexuality in Gay Men and Lesbians

Measure	Gay Men	Lesbians
Genetics	++	+
Sex-typical behavior	+	+
Long bones	+	+
2nd to 4th finger-length ratio	?	+
Finger print ridge counts	-	?
Height	?	?
Weight	-	+
Waist-to-hip ratio	?	+
Age at onset of puberty	+/-	+/-
Penis size	+**	
Otoacoustic emissions	-	+
Auditory evoked potentials	+**	+
Prepulse inhibition	-	+
Smell	+	+
Handedness	+*	+*
Cognitive cerebral lateralization	+	+
Birth order effect	+	-
Maternal behaviors	Stress	Nicotine/drugs

+ Positive findings.
- No positive finding.
? Both positive and negative findings.
*More left-handed.
**Hypermasculinized.

development. Testosterone levels below or above this range appear to lead to hypomasculination (feminization) or hypermasculinization on different measures and are associated with homosexual male development. Thus, the development of homosexuality in men is only partially consistent with the neurohormonal theory. However, the fraternal birth order effect may be one pathway that leads to homosexuality in males that does not appear to be related to lower or higher levels of prenatal testosterone. The relationship between the fraternal birth order effect and prenatal hormone exposure has yet to be characterized.

All of these studies taken together—even accepting that most of them need to be replicated—provide substantial support for the biological origins of human sexual orientation. However, it is important to realize that we do not have definitive proof that sexual orientation in humans is related to exposure to elevated or reduced levels of testosterone. The studies in which testosterone appears to be involved in sexual orientation determination are correlations only. The shear number of such correlations lends support to the likelihood that testosterone is a causal factor in sexual orientation determination, but it is not proof. Much more work will be needed to come to a clearer picture of the role testosterone plays in determining sexual orientation.

What has not been presented is data in support of the social learning theory of sexual orientation determination. This theory says that sexual orientation can be learned after birth through interactions with one's social environment. I would present such data if there were any, but there are no data. If you go to the scientific literature and try to find papers concerning social learning theory and sexual orientation determination, you will find publications that are what I call opinion papers. People discuss current findings and offer opinions, but no data to support their opinions. These papers are useful because they make us think and question, as we should. But if you are a good student of the scientific literature, you recognize that opinions are not data. If an idea is credible, there should be data. Always ask to see the data.

Data does exist on the side of the argument that postnatal social environment is not causally related to sexual orientation determination. A study sponsored by the American Academy of Pediatrics (Pawelski et al., 2006) examined data from the U.S. Census Bureau's 1990 and 2000 census. With regard to gender identity and sexual orientation, no differences were found for over 500 preadolescent children raised by lesbian mothers relative to those raised by heterosexual mothers. No differences were found in the toy, game, activity, dress, or friendship preferences of boys and girls who had lesbian mothers compared with those who had heterosexual mothers. Comparing young adults who had heterosexual mothers to those who had lesbian mothers, similar proportions identified themselves as homosexual.

Thus, it appears that no data indicate that one's sexual orientation is significantly influenced by postnatal social interactions.

So, is one's sexual orientation a choice or is it a matter of biology and therefore nature's choice? Each of you must decide for yourself. At least now you can make an informed decision.

Chapter 16

The Use of Science

Science is a marvelous enterprise. It has given us immense knowledge of the universe, our planet, and us. However, science is a method with which to seek knowledge, truth about what is. We can put no value judgment on knowledge or the truth; there is no good or bad knowledge. How the knowledge obtained through science is used is up to us. Science gave us the knowledge with which to build an atomic bomb, but it was up to us whether to use it or not. The choice was ours. Every future generation will have to choose how to use the awesome power that knowledge has led us to.

Nowhere has science given us more than in biology and medicine. Prior to World War II, bacteria could kill us; people routinely died of pneumonia and other bacterial infections. Today, if treated early enough, almost no one should die of a bacteria infection. This is but one example of science used in the service of humankind, as it should be.

I am often asked whether knowledge of the genetics and general science of homosexuality will lead to an effort to intervene during fetal development to eliminate homosexuals from the general population. There are two questions here. First, should homosexuals be eliminated from the general population? Second, could we, through the knowledge we have gained, achieve that end?

The question of should we attempt to eliminate homosexuals from the general population has an easy answer: no. We are a human family composed of individuals of many diverse variations on a basic plan. Remember, nature tinkers and plays with the expression of genes. Sometimes variations are lethal and the organism doesn't survive. Sometimes the tinkering produces variants whose survival is enhanced. And sometimes the variant survives and is just different.

Nature's Choice

There are no good or bad variations, just some that work better for an organism in a particular habitat. The world at large does not condone the selective elimination of a minority segment of any society. It is a matter of ethics. Take a look at the Web site for the human genome project (http://www.genomics.energy.gov) and you will see that the ethical use of the genome project data is a very serious concern. We are responsible for making sure that this knowledge is not used to harm people. As Carl Sagan said, ". . . the cure for the misuse of science is not censorship, but a clearer explanation, more vigorous debate, and making science accessible to everyone" (Sagan & Druyan, 1992, p. 68). Furthermore, in most Western countries, most particularly in the United States, the rights of the individual, all individuals, are highly protected. These rights include life, liberty, and the pursuit of happiness. We have a marvelous organization called the American Civil Liberties Union that fights to protect the rights of all individuals, including homosexuals.

As to the second part of the question, could we intervene to eliminate homosexuals from the general population? First of all, even if we knew which genes were involved in the development of homosexuality, it is highly likely that those genes are also involved in a process or processes that are used by cells in many different systems throughout development. There is no "gay gene" such that if you turn it on you get homosexuality and if you turn it off you get heterosexuality. Furthermore, we do not know enough to undertake such an intervention. If we did know how to intervene very early in development to alter gene expression throughout development, we might be able to prevent diseases such as Tay-Sachs, Huntington's, and others that are agonizing to watch progress and ultimately lethal. We know the gene involved in both of these diseases and we are helpless to alter the results of having a defective gene, such as in Tay-Sachs, or simply an altered gene whose effects take years to manifest, such as in Huntington's. Second, even if we knew which genes to target, we do not know what to do, or when to do it. Furthermore, the information that we would need to undertake an intervention could come only from experiments on humans, and this can't be done; it would be unethical, and the ethics of scientific work is constantly being scrutinized as it is, as it should be.

Genes aside, what about testosterone? Could we figure out just how much testosterone is needed to provide for heterosexual male development and then make sure that every male fetus was exposed to just that amount? Of course we would be exposing the mother to elevated levels of testosterone as well. There are a large number of variables that we would need to understand and control, but for the sake of discussion, what kind of experiment would we have to conduct? Let's try to determine the levels of testosterone needed and the time during development when they would be required. Let's also say that we have a noninvasive way of making the measurements, which we don't. So, how many pregnant women would we have to identify to participate in our study? Once again, we are looking just at males. We estimate the level of male homosexuality in the general population at about 3 percent. We also estimate that 35 percent of these may owe their sexual orientation to a gene(s) on the X chromosome, which may or may not involve testosterone; so our 3 percent drops to about 1.95 percent. We would control for the fraternal birth order effect by having only women with their first male pregnancy in our study group. In the end, let's say that we would like to have at least 100 gay men in our final sample. This means that we have to start with 5,200 or more pregnant women who are carrying a male fetus. Furthermore, the fraternal birth order effect may not involve reduced levels of testosterone, so we need to take this into account. The fraternal birth order effect accounts for another 0.6 percent, so now we have about 1.35 percent of the males who could be homosexual due to variations in the availability of testosterone. In the end, let's say that we would like to have at least 100 gay men in our final sample. That means that we have to start with 7,400 or more pregnant women who are carrying a male fetus. We would start taking our measurements at the seventh week and continue throughout to birth and perhaps a month or so after birth. We would likely want to take measurements very often, say twice a day through the first six months. Thereafter once a day may suffice. This would be 320 determinations per individual, resulting in a total of 1,660,000 determinations. Then we need to follow all 5,200 individuals until their mid to late teens until they are aware of their sexual orientation and their sexual orientation can be confirmed through assessment on the Kinsey Scale. So, our experiment will involve 5,200 women and their male offspring over a period of 15 to 20 years. This is an enormous undertaking. In practice, following subjects over such a long time is very difficult. Furthermore, the cost of such an undertaking would be enormous and thus prohibitive. The same kind of experiment could be done asking the same question for female offspring. The problem here is that we think we would have to somehow limit or reduce the levels of testosterone available to the fetus. Reducing the level of anything in fetal circulation is something we do not know how to do, nor would we want to make the attempt to do it, as it would be very dangerous for the fetus. Furthermore, steroid hormones, all of them, are very powerful and far-reaching in their effects. Testosterone is particularly dangerous since it can lead to psychosis in high doses. The standard today is to not give steroid hormones or anything that can mimic their actions to pregnant women.

So, should homosexuality be eliminated from the general population? The answer is still no! Could we eliminate homosexuality from the general population? The answer is: not in the foreseeable future. We all need to stay vigilant and informed, so these kinds of experiments are never performed.

Chapter 17

Epilogue

Leaving science behind and at the risk of being too philosophical, I leave you with three quotations and some personal thoughts.

Nothing is harder than not being oneself.

Yvonne Rosnik, Denver PFLAG mom
(personal communication)

Most of us are compelled to be genuine in expressing ourselves. However, if we perceive that some aspect of ourselves is not acceptable to our family, friends, and society, we will hide that part and not openly express it. Hiding parts of ourselves is a painful, constricting process and limits the full expression of our true talents and abilities. It seems to me that one of life's major tasks is discovering who we are in total, and realizing that hiding parts of ourselves leads to considerable psychological stress and unhappiness. This is an ongoing task as we are constantly changing; our personal growth isn't done until we die. The most we can hope for is that all of us are allowed to be ourselves, for only then can we achieve true personal freedom and realize our own unique human potentials. It is in the full expression of our true talents and abilities that we give the most back to the people and world that gave us life.

The privilege of a lifetime is being who you are.

Joseph Campbell (Osborn, 1991, p. 15)

Privilege is defined as a right granted as a benefit, advantage, or favor. How different would life be if we all saw our lives as a privi-

Nature's Choice

lege: something to be grateful for, to honor and cherish. Would we do things differently? Would we put more effort into living life more genuinely? We should all be allowed to embrace our lives as a privilege.

> All true things must change and only that which changes remains true.
>
> Carl Jung (attributed: source unknown)

There is nothing on the planet that isn't constantly changing, including us. We must remain open to what the future will reveal to us about ourselves and the world in which we live. What is revealed is nature's truth. None of us gets to choose who we are, when and where we are born, and the circumstances into which we are born. Similarly, none of us gets to choose whether we are male or female, naturally left- or right-handed, gay or straight. Perhaps, in a way, we are all here to only discover and live nature's truth as it is expressed in each of us the best we can. All of us deserve the right to live that process of discovery with dignity and respect.

Appendix

FERTILIZATION AND GENDER/CHROMOSOME DETERMINATION

All human cells contain 23 pairs of chromosomes. Twenty-two of the chromosome pairs are call autosomes, while the other pair is the sex chromosomes. Fertilization is the union of an egg carrying 22 autosomes plus one X chromosome and a sperm also carrying 22 autosomes, and either an X or Y chromosome (Figure A.1). If a sperm carrying an X chromosome fertilizes an egg, the resultant embryo will have XX chromosomes and be female. Conversely, if a sperm carrying a Y chromosome fertilizes an egg, the embryo will have XY chromosomes and be male. Thus, the father's sperm determines the gender of the embryo, as depicted in Figure A.1

PEDOPHILIA

Pedophilia is defined as a perversion in which children are the preferred sexual object. *Children* means prepubertal individuals, individuals who are not sexually mature. If we accept that sexual attraction is preceded by an emotional response to the object, then pedophiles experience an attractive emotional response to sexually immature individuals. Furthermore, this attraction can be expressed across the same range of sexual orientations as that of adults for whom the sexual object is another adult. Some pedophiles will be heterosexual, some bisexual, and some homosexual. For an individual that molests opposite sex children, the appropriate terminology should be, "heterosexual pedophilic orientation." The other two possible pedophilic orientations are bisexual pedophilic and homosexual pedophilic. Pedophilia is not thought to be curable, but is considered treatable. The individual must learn to control how he or she acts on their impulses.

Nature's Choice

Normal genetics and sex determination

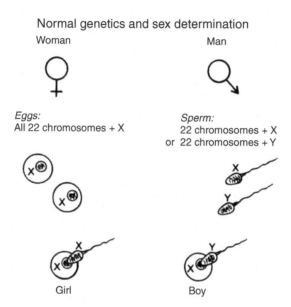

FIGURE A.1. Gender determination at fertilization: The father's sperm determines human gender at fertilization. If the sperm carries an X chromosome, the resultant embryo will be XX, or female. If the sperm carries a Y chromosome, the resultant embryo will be XY, or male. *Source*: Moir & Jessel (1991).

CHILDHOOD SEXUAL ABUSE

The perpetrators of sexual abuse of children are for the most part men. Sexual abuse of children by women most often takes the form of oral stimulation of the adult by the child. Sexual abuse of children by men usually involves either vaginal or anal penetration of the child by the adult.

There is the idea that anal penetration of a boy by an adult male will predispose the boy to prefer anal penetration as a sexual act when he is an adult. Since anal penetration is an act associated with homosexual male sexual behavior, it is assumed that the boy will be homosexual as an adult regardless of his own inherent sexual orientation.

The reverse argument is applied if the child is a girl. In the case of girls, it is thought that if a girl is sexually abused by an adult male (which is usually via vaginal penetration, or conventional heterosexual intercourse) she will turn away from heterosexual intercourse as an adult sexual act. It is thought that this will lead her to prefer lesbian sexual practices regardless of her own inherent sexual orientation.

Taken together, these two arguments defy logic. You can't make the argument one way in the case of boys and reverse it in the case of girls. How can boys be singled out to prefer the abusive sexual act, while girls are presumably repulsed by the abusive sexual act? This does not make sense.

It actually turns out that many male and female prostitutes were sexually abused as children. In part, the psychological explanation for this behavior is "acting out" what was socially learned in childhood. This is an unconscious, nearly reflexive compliance with a historically learned behavior. In the end, childhood sexual abuse appears to impair an individual's ability to express his or her inherent sexuality, including his or her own sexual needs, desires, and orientation. No wonder that many childhood abuse survivors eventually seek psychotherapeutic help as adults in order to deal with difficulties in their intimate relationships.

GENE EXPRESSION AND CELL FUNCTION

In the human condition androgen insensitivity (see also Chapter 4, Figure 4.4), the receptor for testosterone is defective and unable to form the essential steroid-hormone receptor-complex. The result is that all of the cells that are supposed to assume male functions fail to do so and the development of the individual **defaults** to the female path. The process by which the steroid-hormone receptor-complex provides for the production of the proteins required for male functions in cells is shown in Figure A.2. The complex binds to specific areas on chromosomes and provides for transcription of the DNA (gene) sequence to produce RNA within the nucleus. The RNA is modified in a process called splicing to produce mRNA (messenger RNA), which is transported out of the nucleus into the cytosol. In the cytosol, the mRNA associates with large molecular complexes called ribosomes. In the presence of amino acids the mRNA-ribosome complex then produces many copies of a unique protein. Many such proteins are produced, and it is these proteins that enable the cell to carry out its functions according to the male program. For example, facial hair **follicles** grow giving rise to male facial hair. All cells throughout the body that can provide for either male or female functions will provide male functions under the influence of testosterone and related androgens and their receptor complexes.

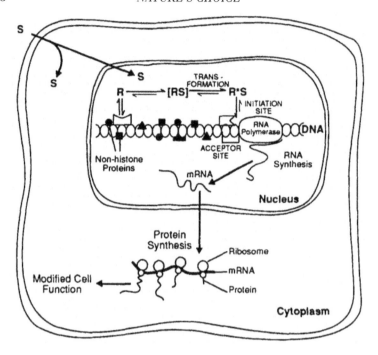

FIGURE A.2. Gene expression: Transcription factors, such as the steroid-hormone receptor-complex, regulate the expression of genes. Genes whose expression is turned on are acted on by enzymes and other molecules in a process call DNA transcription to produce RNA. The induction (or repression) of mRNA synthesis is effected by the action of the enzyme RNA polymerase. The mRNA then enters the cytoplasm where protein synthesis occurs on ribosomes, resulting in modified cell function. *Source:* Adapted from Brown (1994). Reprinted with the permission of Cambridge University Press.

THE SEX CHROMOSOMES

The human sex chromosomes are shown in Figure A.3. All chromosomes have a short arm, denoted "p" (petite) and a long arm, denoted "q." The point where the two arms meet is referred to as the centromere. The distance along the arms is measured from 0, starting at the centromere and measuring out toward the end of either the p or q arm. In the Hamer et al. (1993a, 1993b, 1993c) experiments on linkage of male homosexuality and markers on the X chromosome, linkage was observed at Xq28. Thus, the position on the X chromosome where the linkage was observed was near the end of the long arm (q) at the locus designated 28.

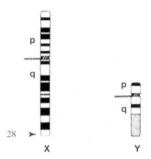

28

X   Y

FIGURE A.3. The X chromosome: The human X chromosome showing the area Xq28, thought to be linked to some male homosexuality. The Y chromosome is shown for comparative purposes. *Source:* Thompson, McInnes, & Willard (1991). Reprinted from *Thompson & Thompson Genetics in Medicine,* M.W. Thompson, R.R. McInnes, and H.F. Willard, copyright 1991, with permission from Elsevier.

Glossary

activation effect: An effect of a hormone on brain or body function in adulthood that lasts only as long as the hormone is present.

androgens: A class of gonadal steroids, including testosterone, whose typical hormonal action is to drive development or differentiation in the male direction.

anterior commissure: A bundle of axons, or nerve fibers, connecting the left and right hemispheres of the cerebral cortex of the middle and inferior temporal lobes. It is similar in function to the corpus callosum, but much smaller, and it mediates the interhemispheric transfer of visual, auditory, and olfactory information.

aromatase: An enzyme that transforms androgens, specifically testosterone, into estrogen.

axon: One of two types of structural projections from a neuron that carries information to the neuron's target cell.

chromosome: A discrete unit of a genome, consisting of a very long molecule of DNA, containing many genes in a specific sequence.

clitoris: A structural component of the female external genitalia; a protuberance lying in the front of the urethra where the labia minora fuse at the midline. It is an erectile organ and a major site of sexual excitability.

cognitive: Pertaining to the mental basis of knowledge and perception. The word is increasingly applied to mental processes in general.

congenital adrenal hyperplasia: A genetic abnormality of steroid hormone biosynthesis that leads to excess secretion of androgens by the adrenal glands during fetal life, and hence can lead to partial masculinization of female fetuses.

Nature's Choice

corpus callosum: The main axonal pathway interconnecting the left and right hemispheres of the cerebral cortex.

cortex: Also neocortex, is the sheetlike expanse of gray matter on the surface of the brain and made up of several layers of neurons.

corticosteroids: A group of steroid hormones produced by the adrenal gland that do not play an active role in the regulation of sexual development or behavior.

default: In the absence of a specific instruction or signal.

differentiation: The sum of the processes whereby apparently indifferent cells, tissues, and structures attain their adult form and function.

dihydrotestosterone (DHT): An androgen, more potent than testosterone, synthesized from testosterone by an enzyme in certain tissues.

dimorphic: Occurring in two distinct forms.

DNA: Deoxyribonucleic acid, the molecular components of heredity.

enzyme: A protein that catalyzes (promotes) a specific biochemical reaction.

epigenetic: Any factor that can affect the phenotype without a change in geneotype.

estradiol: The major estrogenic steroid.

estrogens: A class of gonadal steroids produced by granulosa cells of the ovarian follicles. Estrogens are responsible for the completion of female development at puberty, and play a key role in the menstrual cycle and pregnancy. In males, estrogens are also produced by the testes and synthesized from androgens by some brain cells, playing a role in the male sexual development of the brain.

estrous cycle: The sexual cycle in female animals, corresponding to the menstrual cycle in women. It is generally shorter than in humans and marked by major changes in sexual receptivity.

female-typical: Referring to behavior, those behaviors seen more commonly in females than males.

follicle: The developing oocyte with its surrounding granulosa and thecal cells.

gamete: Either type of reproductive cell, the male sperm and female ovum or egg, that mixes its genes in sexual reproduction.

gene: A unit of heredity, consisting of a segment of DNA involved in producing a polypeptide chain or protein, several regulatory sequences, and some additional noncoding sequences.

gene expression: The activity of a gene, generally meaning the production of the protein for which it codes.

genitalia: Sex organs, usually excluding the gonads. These are grouped into *internal genitalia* (oviducts, uterus, and cervix in females; epididymis, vas deferens, prostate gland, and seminal vesicles in males) and the *external genitalia* (vagina, labia minora, labia majora, and clitoris in females; penis and scrotum in males).

genome: The total genetic information stored in the chromosomes of an organism.

gonad: The organ that produces gametes—the testes in males and the ovaries in females.

gonadal steroids: Steroids produced by the gonads (testis and/or ovary); sex hormones.

granulosa cells: Cells surrounding the developing ovum that are involved in the production of estrogens.

gray matter: The parts of the brain and spinal cord that contain neurons and synapses; it takes the form of either cortex or nuclei.

heterosexual: Having sexual feelings for or behavior directed toward individuals of the opposite sex.

heterozygous: Having different versions of a gene on the two members of a pair of chromosomes.

homosexual: Having sexual feelings for or behavior directed toward individuals of one's own sex.

homozygous: Having the same version of a gene on both members of a pair of chromosomes.

hormone: A chemical or protein secreted by specialized endocrine or neuroendocrine cells that is transported, usually via the blood stream, to other cells or tissues, where it influences their metabolism or development.

hypothalamus: A small region at the base of the brain containing a number of nuclei or areas involved in the regulation of instinctual drives, cardiopulmonary function, endocrine function, metabolism, food and water balance, and temperature regulation.

INAH 1-4: The four interstitial nuclei of the anterior hypothalamus. INAH 3 is dimorphic with sex or gender and sexual orientation; INAH 2 may be dimorphic with sex or gender; INAH 1 (also known as the nucleus intermedius) and INAH 4 are probably not dimorphic with either sex (gender) or sexual orientation.

intromission: The insertion of the penis into an orifice such as the vagina.

labia majora: The tissue folds surrounding the labia minora.

labia minora: The inner tissue folds surrounding the vagina and urethra.

Leydig cells: Androgen-producing cells of the testis.

linkage: Genes on the same chromosome show linkage if they have a tendency to be transmitted together through meiosis, or the type of cell division that takes place in gametes (sperm or ovum).

lordosis: In rodents and some other mammals, a reflex curving of the back into a U-shape that exposes the genitalia and permits intromission.

luteinizing hormone (LH): A peptide hormone secreted by the pituitary gland that induces the final maturation of the developing oocyte in females. In males it stimulates testosterone production by the Leydig cells of the testis.

luteinizing hormone releasing hormone (LHRH): A peptide hormone synthesized by a special set of hypothalamic neurons that is transported by blood vessels to the pituitary gland, where it regulates the secretion of luteinizing hormone.

male-typical: Seen more commonly in males than females.

medial: Toward the midline.

medial preoptic area: The portion of the preoptic area closest to the midline.

menstrual cycle: The cycle (monthly in humans) in which the uterus alternates between a state suitable for the transport of sperm and a state suitable for the implantation of the embryo.

Müllerian duct: The embryonic precursor of the female internal genitalia.

Müllerian inhibition hormone (MIH): A hormone produced by the male gonad during early development that prevents the Müllerian duct from developing into the internal female genital tract.

mutation: A change in a gene.

neuron: A nerve cell.

nucleus (plural nuclei): 1. the central compartment of a cell, which contains most of its genetic material. 2. in neuroanatomy, a coherent assembly of neurons recognizable as a discrete structure in the brain; the neurons in a nucleus are generally similar to each other in structure, chemistry, connections, and function.

nucleus intermedius: A nucleus in the human hypothalamus, also known as INAH 1. Whether it is sexually dimorphic is presently unclear.

olfactory: To smell; to have a smell; of the sense of smell

optic chiasm: The crossing of the optic nerves—a landmark at the front of the hypothalamus.

organizational effect: A feature of brain organization that is brought about by hormonal exposure during early development, and that permanently influences an individual's behavior.

ovary: The female gonad, a paired organ within the pelvic cavity that produces ova and sex hormones.

ovulation: The release of an unfertilized ovum from the ovary.

ovum (plural ova): An egg, before or after fertilization. Before fertilization it is a female gamete or oocyte; after fertilization it is a zygote.

pedigree analysis: Tracing a trait through several generations of a family using a family tree.

penis: The major component of the male external genitalia, consisting of glans (head) and shaft. The penis contains erectile tissue and is traversed by the urethra. It serves for voiding urine, sexual arousal, intromission, and ejaculation of semen. It is derived from the same tissue as the clitoris and the labia minora in females.

PET scan: A technique that combines the principles of computerized tomography, with radioisotope imaging for producing images of single planes of the live brain during real-time brain functioning. It is a noninvasive imaging technique for visualizing the local changes in cerebral blood flow and metabolism that accompany mental activities.

phenotype: The physical expression of genes; also the trait or traits created by gene expression.

pheromone: Volatile compounds (molecules) secreted into the environment (in sweat or urine) by one individual of a species, perceived by another individual of the same species, in whom they trigger a behavioral response or physiological change.

pituitary gland: A gland connected by a stalk to the undersurface of the hypothalamus. It secretes a large number of hormones that regulate the function of other glandular organs, including the gonads and mammary glands.

preoptic area: The foremost portion of the hypothalamus, from the optic chiasm forward.

presenting: Female-typical sexual behavior in lower primates and mammals—the display of the genital area as a stimulus to mounting. It can be either proceptive (spontaneous) or receptive (provoked by a partner's behavior).

proceptive behavior: Female-typical behavior that actively solicits sex, for example, staring or spontaneous presenting in monkeys, ear-wiggling in rats (contrast with *receptive behavior*).

progesterone: Principal member of a class of steroid hormones (progestins) produced by the ovary that plays a role in the menstrual cycle and the maintenance of pregnancy.

puberty: The transition from childhood to sexual maturity.

receptive behavior: Female-typical sexual behavior (lordosis in rats) that is shown in response to the partner's behavior (mounting by a male rat) and that permits copulation (contrast with *proceptive behavior*).

receptor: Usually a large protein molecule, whose chemical structure allows it to recognize and bind a particular hormone, neurotransmitter, or other small molecule.

scrotum: Component of the male external genitalia—the sac that encloses the testes. It derives from the same tissue as the labia majora in females.

semen: The fluid ejaculated from the penis at orgasm; it contains sperm along with the secretions of the prostate and other glands.

seminal emission: The discharge of semen into the urethra, usually followed immediately by ejaculation.

sex: An individual's maleness or femaleness, based primarily on the anatomy of his or her external genitalia.

sex chromosomes: The pair of chromosomes (X or Y) responsible for sex determination.

sex-atypical: More common in the opposite sex.

sex-linked: Inherited along with sex by means of genes located on the sex chromosomes.

sexual differentiation: The process by which the fetus becomes internally and externally male or female.

sexual dimorphism: An anatomical difference between the sexes.

sexual orientation: The direction of sexual feelings and behavior toward the same sex, opposite sex, or some combination of the two.

sexual reproduction: A form of reproduction involving the merging of genetic material from two individuals.

sexually dimorphic nucleus: A nucleus in the medial preoptic area of the rat's hypothalamus that is larger in males than in females.

significance: The statistical assessment of how likely it is that a scientific finding occurred simply by chance.

sperm (or spermatozoa): The male gametes—the cellular constituents of semen.

splenium: The enlarged back end of the corpus callosum.

steroids: A class of organic molecules synthesized from cholesterol.

suprachiasmatic nucleus: A nucleus of the hypothalamus that regulates the daily cycles of activity, temperature, and sleep.

supraoptic nucleus: One of two hypothalamic nuclei that regulate milk ejection, uterine contraction, and urine flow.

synapse: A point of contact between two neurons or a neuron and its target cell where the activity of a neuron excites or inhibits the activity of another neuron or target cell.

testes (singular testis) or testicle: The male gonad, a paired organ lying within the scrotum that produces sperm and sex hormones.

testis-determining factor (TDF): The sex-determining gene in mammals, located on the Y chromosome. It confers maleness by causing the gonads to develop as testes.

testosterone: The primary male steroid-hormone.

thalamus: A large group of nuclei deep within the cerebral hemispheres. The thalamus controls the inputs to the cerebral cortex and modulates the activity level of the cortex.

thecal cells: The outer cells of the ovarian follicle. They synthesize androgens that are subsequently converted into estrogens by the granulosa cells.

third ventricle: A slit-shaped ventricle, or cavity, in the midline of the brain that divides the thalamus and hypothalamus into left and right halves.

tract: An axonal pathway.

transsexuality: Discordance between sex and gender identity.

urethra: The tube through which urine is conveyed from the bladder to the exterior of the body. In males, ejaculation of semen also takes place through the urethra.

uterine contiguity effect: The partial masculinization of female rodent fetuses by exposure to testosterone secreted by nearby male fetuses.

uterus: The womb; site of implantation and development of the embryo.

vagina: The sheathlike component of the female external genitalia that forms a passage between the uterus and the exterior. It serves for sexual arousal, penile intromission, passage of sperm and menstrual fluid, and as the last portion of the birth canal.

ventricle: A cavity within the brain filled with cerebrospinal fluid.

ventromedial: A hypothalamic nucleus involved in the generation of female-typical sexual behavior, as well as other functions including the regulation of appetite.

white matter: The portions of the brain and spinal cord that contains only axons or nerve fibers, not neuronal cell bodies or synapses.

Wolffian duct: The embryonic precursor of the male internal genitalia.

X-linkage: Genes on the X chromosome, or traits determined by such genes, are X-linked.

Xq28: The region at position 28 on the long arm (q) of the X-chromosome. The location of gene linkage determined by Dean Hamer to be associated with male homosexual orientation in some men.

zygote: The fertilized ovum—the single cell, formed by fusion of male and female gametes, that divides repeatedly to form an embryo.

Recommended Reading

FURTHER READING

These are general presentations of the differences between males and females across a broad range of behaviors and how sex contributes to those differences. References to the biology of sexual orientation are included.

1. Moir, A. & Jessel, D. (1991). *Brain Sex.* New York: Dell Publishing.
2. Blum, D. (1997). *Sex on the Brain.* New York: Penguin Putnam.

This article presents detailed descriptions of sexual development, androgen insensitivity, and 5-alpha reductase deficiency.

3. Diamond, J. (1992). Turning a Man. *Discover, 3,* 70-77.

This is a more scientifically based description of how the brain contributes to the observed differences between males and females. It includes references to the biology and brain structures that are associated with sexual orientation.

4. Le Vay, S. (1993). *The Sexual Brain.* Cambridge, MA: Bradford Books.

The personal story of how the linkage of some male homosexuality with the X chromosome was identified as told by Dean Hamer.

5. Hamer, D. & Copeland, P. (1994). *The Science of Desire.* New York: Simon and Schuster.

This the most recent research-base presentation of the areas of human biology that have been examined with reference to homosexuality.

6. Wilson, G. & Rahman, Q. (2005). *Born Gay: The Psychobiology of Sex Orientation.* London: Peter Owens.

Homosexuality examined from a number of perspectives, including science, history, and politics.

Nature's Choice

7. Burr, C. (1996). *A Separate Creation: The Search for the Biological Origins of Sexual Orientation.* New York: Hyperion.

This is an excellent general college-level text of neurobiology and behavior.

8. Kandel, E.R., Schwartz, J.H., & Jessel, T.M. (1995). *Essentials of Neural Science and Behavior.* Stamford, CT: Appleton & Lange.

Bibliography

Alexander, J.E. & Sufka, K.J. (1993). Cerebral lateralization in homosexual males: A preliminary EEG investigation. *International Journal of Psychophysiology, 15*(3), 269-274.

Allen, L.S. & Gorski, R.A. (1992). Sexual orientation and the size of the anterior commissure in the human brain. *Proceedings of the National Academy of Sciences, USA, 89I*(15), 7199-7202.

Allen, L.S., Hines, M., Shryne, J.E., & Gorski, R.A. (1989). Two sexually dimorphic cell groups in the human brain. *Journal of Neuroscience, 9*(2), 497-506.

American Psychiatric Association (2000). COPP position statement on therapies focused on attempts to change sexual orientation (reparative or conversion therapies). Retrieved from: http://www.psych.org/psych_pract/copptherapyaddendums 83100.cfm?pf=y.

American Psychological Association (1997). Resolution on appropriate therapeutic responses to sexual orientation. Retreived from: http://www.apa.org/pi/sexual .html.

Annett, M. (1985). *Left, right, hand and brain: The right shift theory.* Hillsdale, NJ: Erlbaum.

Arnold, A.P. (1996). Genetically triggered sexual differentiation of the brain and behavior. *Hormones Behavior, 30*(4), 495-505.

Arnold, A.P. & Breedlove, S.M. (1985). Organizational and activational effects of sex steroids on the brain and behavior: A reanalysis. *Hormones and Behavior, 19*(4), 469-498.

Arnold, A.P. & Gorski, R.A. (1984). Gonadal steroid induction of structural sex differences in the central nervous system. *Annual Review of Neuroscience, 7,* 413-442.

Bagemihl, B. (1999). *Biological exuberance: Animal homosexuality and natural diversity.* London: Profile Books, Ltd.

Bailey, J.M. & Bell, A.P. (1993). Familiality of female and male homosexuality. *Behavior Genetics, 23*(4), 313-322.

Bailey, J.M. & Benishay, D.S. (1993). Familial aggregation of female sexual orientation. *American Journal of Psychiatry, 150*(2), 272-277.

Bailey, J.M., Dunne, M.P., & Martin, N.G. (2000). Genetic and environmental influences on sexual orientation and its correlates in an Australian twin sample. *Journal of Personality and Social Psychology, 78*(3), 524-536.

Nature's Choice

Bailey, J.M. & Pillard, R.C. (1991). A genetic study of male sexual orientation. *Archives of General Psychiatry, 48*(12), 1089-1096.

Bailey, J.M. & Pillard, R.C. (1995). Genetics of human sexual orientation. *Annual Review of Sex Research, 6,* 126-150.

Bailey, J.M., Pillard, R.C., Dawood, K., Miller, M.B., Farrer, L.A., & Trivedi, S., et al. (1999). A family history study of male sexual orientation using three independent samples. *Behavior Genetics, 29*(2), 79-86.

Bailey, J.M., Pillard, R.C., Neale, M.C., & Agyei, Y. (1993). Heritable factors influence sexual orientation in women. *Archives of General Psychiatry, 50*(3), 217-223.

Bailey, J.M., Willerman, L., & Parks, C. (1991). A test of the maternal stress theory of human male homosexuality. *Archives of Sexual Behavior, 20*(3), 277-293.

Bailey, J.M. & Zucker, K.J. (1995). Childhood sex-typed behavior and sexual orientation: A conceptual analysis and quantitative review. *Developmental Psychology, 31*(1), 43-55.

Bardin, C.W. (1983). The androgenic, antiandrogenic, and synandrogenic actions of progestins. In C. W. Bardin, E. Milgrom, and P. Mauvais-Jarvis (Eds.), *Progesterone and progestins* (pp. 135-147). New York: Raven Press.

Beach, F.A. (1975). Hormonal modifications of sexually dimorphic behavior. *Psychoneuroendocrinology, 1*(1), 3-8.

Bell, A.P., Weinberg, M.S., & Hammersmith, S.K. (1981). *Sexual preference: Its development in men and women.* Bloomington: Indiana University Press.

Berglund, H., Lindstrom, P., & Savic, I. (2006). Brain responses to putative pheromones in lesbian women. *Proceedings of the National Academy of Sciences, USA, 103*(21), 8269-8274.

Blanchard, R. (1997). Birth order and sibling sex ratio in homosexual versus heterosexual males and females. *Annual Review of Sex Research, 8,* 27-67.

Blanchard, R. (2001). Fraternal birth order and the maternal immune hypothesis of male homosexuality. *Hormones and Behavior, 40*(2), 105-114.

Blanchard, R. (2004). Quantitative and theoretical analyses of the relation between older brothers and homosexuality in men. *Journal of Theoretical Biology, 230*(2), 173-187.

Blanchard, R. (2007). Sex ratio of older siblings in heterosexual and homosexual, right-handed and non-right-handed men. *Archives of Sexual Behavior.* DOI: 10.1007/s10508-006-9119-2.

Blanchard, R. & Bogaert, A.F. (1996a). Biodemographic comparisons of homosexual and heterosexual men in the Kinsey interview data. *Archives of Sexual Behavior, 25*(6), 551-579.

Blanchard, R. & Bogaert, A.F. (1996b). Homosexuality in men and number of older brothers. *American Journal of Psychiatry, 153*(1), 27-31.

Blanchard, R. & Bogaert, A.F. (2004). Proportion of homosexual men who owe their sexual orientation to fraternal birth order: An estimate based on two national probability samples. *American Journal of Human Biology, 16*(2), 151-157.

Blanchard, R., Cantor, J.M., Bogaert, A.F., Breedlove, S.M., & Ellis, L. (2006). Interaction of fraternal birth order and handedness in the development of male homosexuality. *Hormones and Behavior, 49*(3), 405-414.

Blanchard, R., Dickey, R., & Jones, C.L. (1995). Comparison of height and weight in homosexual versus nonhomosexaul gender dysphorics. *Archives of Sexual Behavior, 24*(5), 543-554.

Blanchard, R. & Ellis, L. (2001). Birth weight, sexual orientation, and the sex of preceding siblings. *Journal of Biosocial Science, 33*(3), 451-467.

Blanchard, R. & Lippa, R.A. (2006a). Birth order, sibling sex ratio, handedness, and sexual orientation of male and female participants in a BBC Internet research project. *Archives of Sexual Behavior, 36*(2), 163-176.

Blanchard, R. & Lippa, R.A. (2006b). The sex ratio of older siblings in non-right-handed homosexual men. *Archives of Sexual Behavior, 26,* 163-176.

Blanchard, R., Zucker, K.J., Cavacas, A., Allin, S., Bradley, S.J., & Schachter, D.C. (2002). Fraternal birth order and birth weight in probably prehomosexual feminine boys. *Hormones and Behavior, 41*(3), 321-327.

Bocklandt, S., Horvath, S., Vilain, E., & Hamer, D.H. (2006). Extreme skewing of X cromosome inactivation in mothers of homosexual men. *Human Genetics, 118*(6), 691-694.

Boehmer, U., Bowen, D.J., & Bauer, G.R. (2007). Overweight and obesity in sexual-minority women: Evidence from population-based data. *American Journal of Public Health, 97*(6), 1134-1140.

Bogaert, A.F. (1998a). Birth order and sibling sex ratio in homosexual and heterosexual non-white men. *Archives of Sexual Behavior, 27*(5), 467-473.

Bogaert, A.F. (1998b). Physical development and sexual orientation in women: Height, weight and age of puberty comparisons. *Personality and Individual Differences, 24*(1), 115-121.

Bogaert, A.F. (2003a). The interaction of fraternal birth order and body size in male sexual orientation. *Behavioral Neuroscience, 117*(2), 381-384.

Bogaert, A.F. (2003b). Interaction of older brothers and sex-typing in the prediction of sexual orientation in men. *Archives of Sexual Behavior, 32*(2), 129-134.

Bogaert, A.F. (2003c). Number of older brothers and sexual orientation: New tests and the attraction/behavior distinction in two national probability samples. *Journal of Personality and Social Psychology, 84*(3), 644-652.

Bogaert, A.F. (2005). Gender role/identity and sibling sex ratio in homosexual men. *Journal of Sex & Marital Therapy, 31*(3), 217-227.

Bogaert, A.F. (2006). Biological versus nonbiological oler brothers and men's sexual orientation. *Proceedings of the National Academy of Sciences, USA, 103*(28), 10771-10774.

Bogaert, A.F. & Blanchard, R. (1996). Physical development and sexual orientation in men: Height, weight and age of puberty differences. *Personality and Individual Differences, 21*(1), 77-84.

Bogaert, A.F. & Cairney, J. (2004). The interaction of birth order and parental age on sexual orientation: An examination in two samples. *Journal of Biosocial Science, 36*(1), 19-37.

Bogaert, A.F. & Friesen, C. (2002). Sexual orientation and height, weight, and age of puberty: New tests from a British national probability sample. *Biological Psychology, 59*(2), 135-145.

Bogaert, A.F., Friesen, C., & Klentrou, P. (2002). Age of puberty and sexual orientation in a national probability sample. *Archives of Sexual Behavior, 31*(1), 73-81.

Bogaert, A.F. & Hershberger, S. (1999). The relation between sexual orientation and penile size. *Archives of Sexual Behavior, 28*(3), 213-221.

Bogaert, A.F. & Liu, J. (2006). Birth order and sexual orientation in men: Evidence for two independent interactions. *Journal of Biosocial Science, 38*(6), 811-819.

Bouchard, T.J., Lykken, D.T., McGue, M., Segal, N.L., & Tellegen, A. (1990). Sources of human psychological differences: The Minnesota study of twins reared apart. *Science, 250*(4978), 223-228.

Braff, D.L., Stone, C., Callaway, E., Geyer, M.A., Glick, I., & Bali, L. (1978). Prestimulus effects on human startle reflex in normals and schizophrenics. *Psychophysiology, 15*(4), 339-343.

Brown, R.E. (1994). *An introduction to endocrinology.* Cambridge, England: Cambridge University Press.

Brown, W.M., Finn, C.J., & Breedlove, S.M. (2001). A sex difference in digit length ratio in mice (abstract). *Hormones and Behavior, 39,* 325.

Brown, W.M., Finn, C.J., Cooke, B.M., & Breedlove, S.M. (2002). Differences in finger length ratio between self-identified "butch" and "femme" lesbians. *Archives of Sexual Behavior, 31*(1), 123-127.

Brown, W.M., Hines, M., Fane, B.A., & Breedlove, M.S. (2002). Masculinized finger length patterns in human males and females with congenital adrenal hyperplasia. *Hormones and Behavior, 42*(4), 380-386.

Buhrich, N., Bailey, J.M., & Martin, N.G. (1991). Sexual orientation, sexual identity, and sex-dimorphic behavior in male twins. *Behavior Genetics, 21*(1), 75-96.

Burr, C. (1996). *A separate creation.* New York: Hyperion.

Byne, W. (1994). The biological evidence challenged. *Scientific American, 270*(5), 50-55.

Byne, W., Lasco, M.S., Kemether, E., Shinwari, A., Edgar, M.A., Margello, S., Jones, L.B., & Tobet, S. (2000). The interstitial nuclei of the human anterior hypothalamus: An investigation of sexual variation in volume and cell size, number and density. *Brain Research, 856*(1-2), 254-258.

Byne, W., Tobet, S., Mattiace, L., Lasco, M.S., Kemether, E., Edgar, M.A., Margello, S., Buchsbaum, M.S., & Jones, L.B. (2001). The interstitial nuclei of the anterior hypothalamus: An investigation of variation within sex, sexual orientation and HIV status. *Hormones and Behavior, 40*(2), 86-92.

Cadenhead, K.S., Carasso, B.S., Swerdlow, N.R., Geyer, M.A., & Braff, D.L. (1999). Pre-pulse inhibition and habituation of the startle response are stable neurobiological measures in a normal male population. *Biological Psychiatry,* *45*(3), 360-364.

Cantor, J.M., Blanchard, R., Paterson, A.D., & Bogaert, A.F. (2002). How many gay men owe their sexual orientation to fraternal birth order? *Archives of Sexual Behavior, 31*(1), 63-71.

Carlson, B.M. (1994). *Human embryology and developmental biology.* St. Louis, MO: Mosby.

Choi, J. & Silverman, I. (1996). Sexual dimorphism in spatial behaviors: Applications to route learning. *Evolution and Cognition, 2,* 165-171.

Cole-Harding, S., Morstad, A.L., & Wilson, J.R. (1988). Spatial ability in members of opposite-sex twin pairs (abstract). *Behavior Genetics, 18*(6), 710.

Collaer, M.L., Reimers, S., & Manning, J.T. (2007). Visuospatial performance on an internet line judgment task and potential hormonal markers: Sex, sexual orientation, and 2D:4D. *Archives of Sexual Behavior, 36*(2), 177-192.

Crews, D. (1994). Animal sexuality. *Scientific American, 270*(1), 109-114.

Dabbs, J.M., Chang, E.L., Strong, R.A., & Milun, R. (1998). Spatial ability, navigation strategy, and geographic knowledge among men and women. *Evolution and Human Behavior, 19,* 89-98.

Diamond, J. (1992). Turning a man. *Discover, 13*(2), 71-77.

Diamond, J. (1997). *Why is sex fun? The evolution of human sexuality.* New York: Basic Books.

Diamond, M.C. (1984). Age, sex, and environmental influences on anatomical asymmetry in the rat forebrain. In N. Geschwind & A.M. Galaburda (Eds.), *Cerebral dominance: The biological foundations* (pp. 134). Cambridge, MA: Harvard University Press.

Diamond, M.C., Johnson, R.W., & Ingham, C.A. (1975). Morphological changes in the young, adult, and aging rat cerebral cortex, hippocampus and diencephalons. *Behavioral Biology, 14*(2), 163-174.

Dittmann R.W., Kappes M.E., & Kappes, M.H. (1992). Sexual behavior in adolescent and adult females with congenital adrenal hyperplasia. *Psychoneuroendocrinology, 17*(2-3), 153-170.

Dittmar, M. (1998). Finger ridge-count asymmetry and diversity of Andean Indians and inter-population comparisons. *American Journal of Physical Anthropology, 105*(3), 377-393.

Dorner, G., Greier, T., Ahrens, L., Krell, L., Munx, G., Sieler, H., Kittner, E., & Muller, H. (1980). Prenatal stress possible aetiogenic factor homosexuality in human male. *Endokrinologie, 75*(3), 365-368.

Dorner, G., Schenk, B., Schmiedel, B., & Ahrens, L. (1983). Stressful events in prenatal life bi- and homosexual men. *Experimental Clinical Endocrinology, 81*(1), 83-87.

DuPree, M.G., Mustanski, B.S., Bocklandt, S., Nievergelt, C., & Hamer, D.H. (2004). A candidate gene study of CYP19 (aromatase) and male sexual orientation. *Behavioral Genetics, 34*(3), 243-250.

Eckert, E.D., Bouchard, T.J., Bohlen, J., & Heston, L.L. (1986). Homosexuality in monoztgotic twins reared apart. *British Journal of Psychiatry, 148,* 421-425.

Ellis, L. (1996). The role of perinatal factors in determining sexual orientation. In R.C. Savin-Williams & K.M. Cohen (Eds.), *The lives of lesbians, gays, and bisexuals: Children to adults* (pp. 35-70). New York: Harcourt Brace.

Ellis, L. & Ames, M.A. (1987). Neurohormonal functioning and sexual orientation: A theory of homosexuality-heterosexuality. *Psychological Bulletin, 101*(2), 233-258.

Ellis, L., Ames, M.A., Peckham, W., & Burke, D. (1998). Sexual orientation of human offspring may be altered by severe maternal stress during pregnancy. *Journal of Sex Research, 25*(1), 152-157.

Ellis, L. & Cole-Harding, S. (2001). The effects of prenatal stress, and of prenatal alcohol and nicotine exposure, on human sexual orientation. *Physiology and Behavior, 74*(1-2), 213-226.

Ellis, L. & Hellberg, J. (2005). Fetal exposure to prescription drugs and adult sexual orientation. *Personality and Individual Differences, 38*(1), 225-236.

Ellis, L., Robb, B., & Burke, D. (2005). Sexual orientation in United States and Canadian college students. *Archives of Sexual Behavior, 34*(5), 569-581.

Evens, R.B. (1972). Physical and biochemical characteristics of homosexual men. *Journal of Consulting and Clinical Psychology, 39*(1), 140-147.

Faber, K.A. & Hughes Jr., C.L. (1992). Anogenital distance at birth as a predictor of volume of the sexually dimorphic nucleus of the preoptic area of the hypothalamus and pituitary responsiveness in castrated adult rats. *Biology of Reproduction, 46*(1), 101-104.

Faraday, M.M., O'Donoghue, V.A., & Grumberg, N.E. (1999). Effects of nicotine and stress on startle amplitude and sensory gating depend on rat strain and sex. *Pharmacology, Biochemistry and Behavior, 62*(2), 273-284.

Fausto-Sterling, A. & Balaban, E. (1993). Genetics and male sexual orientation. *Science, 261*(5126), 1257-1259.

Forastieri, V., Andrade, C.P., Souza, A.L., Silva, M.S., El-Hani, C.N., Moreira, L.M., Mott, L.R., & Flores, R.Z. (2002). Evidence against a relationship between dermatoglyphic asymmetry and male sexual orientation. *Human Biology, 74*(6), 861-870.

Garn, S.M., Burdi, A.R., Babler, W.J., & Stinson, S. (1975). Early prenatal attainment of adult metacarpal-phalangeal rankings and proportions. *American Journal of Physical Anthropology, 43*(3), 327-332.

Gebhard, P.H. & Johnson, A.B. (1979). *The Kinsey Data: Marginal tabulations of the 1938-1963 interviews conducted by the Institute for Sex Research.* Philadelphia: W.B. Saunders.

George, R. (1930). Human finger types. *Anatomical Record, 46,* 199-204.

Gladue, B.A. & Bailey, J.M. (1995). Spatial ability, handedness and human sexual orientation. *Psychoneuroendocrinology, 20*(5), 487-497.

Glickman, S.E., Frank, L.G., Davidson, J.M., Smith, E.R. & Siiteri, P.K. (1978). Androstenedione may organize or activate sex-reversed traits in female spotted hyaenas. *Proceedings of the National Academy of Sciences, USA, 84*(10), 3444-3447.

Golombok, S. & Rust, J. (1993a). The measurement of gender role behavior in pre-school behavior. A research note. *Journal of Child Psychology and Psychiatry, 34*(5), 805-811.

Golombok, S. & Rust, J. (1993b). The Pre-School Activities Inventory. A standardized assessment of gender role in children. *Psychological Assessment, 5*(2), 131-136.

Goy, R.W., Bercovitch, F.B., & McBrair, M.C. (1988). Behavioral masculinization is independent of genital masculinization in prenatally androgenized female rhesus macques. *Hormones and Behavior, 22*(4), 552-571.

Goy, R.W. & McEwen (1980). *Sexual differentiation of the brain*. Cambridge, MA: MIT Press.

Goy, R.W. & Phoenix, C.H. (1971). The effects of testosterone proprionate administered before birth on the development of behavior in genetic female Rhesus monkeys. In C. Sawyer & R. Gorski (Eds.), *Steroid hormones and brain function* (pp. 193-201). Berkeley, CA: University of California Press.

Graham, F.K. (1975). The more or less startling effects of weak pre-stimuli. *Psychophysiology, 12,* 238-248.

Grimshaw, G.M., Bryden, M.P., & Finegan, J.-A. (1995). Relations between prenatal testosterone and cerebral lateralization in children. *Neuropsychology 9*(1), 68-70.

Grumbach, M.M. & Conte, F.A. (1998). Disorders of sex differentiation. In J.D. Wilson, D.W. Foster, H.M. Kronenberg, and P.R. Larsen, (Eds.), *Williams textbook of endocrinology* (pp. 1303-1425). Philadelphia: W.B. Saunders.

Hafez, E.S.E. (Ed.) (1993). *Reproduction in farm animals*. Philadelphia: Lea and Febiger.

Hall, J.A. & Kimura, D. (1994). Dermatoglyphic asymmetry and sexual orientation in men. *Behavioral Neuroscience, 108*(6), 1203-1206.

Hall, L.S. (2000a). Dermatoglyphic analysis of monozygotic twins discordant for sexual orientation. In N.M. Durham, K.M. Fox, and C.C. Plato (Eds.), *The state of dematoglyphics: The science of finger and palm prints* (Vol. 2, pp. 123-165). Lewiston, NY: Edwin Mellen Press.

Hall, L.S. (2000b). Dematoglyphic analysis of total ridge count in female monozygotic twins discordant for sexual orientation. *Journal of Sex Research, 37*(4), 315-320.

Hall, L.S. & Love, C.T. (2003). Finger-length ratio in female monozygotic twins discordant for sexual orientation. *Archives of Sexual Behavior, 32*(1), 23-28.

Hamer, D.H. (1999). Technical comment: Genetics and male sexual orientation. *Science, 285*(5429), 803a.

Hamer, D.H., Hu, S., Magnuson, V., Hu, N., & Pattatucci, A.M.L. (1993a). Genetics and male sexual orientation. Response. *Science, 261*(5119), 1257-1259.

Hamer, D.H., Hu, S., Magnuson, V., Hu, N., & Pattatucci, A.M.L. (1993b). A linkage between DNA markers on the X chromosome and male sexual orientation. *Science, 261*(5119), 321-327.

Hamer, D.H., Hu, S., Magnuson, V., Hu, N., & Pattatucci, A.M.L. (1993c). Male sexual orientation and genetic evidence. Response. *Science, 261*(5119), 2063-2065.

Hampson, E., Rovet, J.F., & Altmann, D. (1998). Spatial reasoning in children with congenital adrenal hyperplasia due to 21-hydroxylase deficiency. *Developmental Neuropsychology, 14,* 299-320.

Hare, E.H. & Price, J.S. (1969). Birth order and family size: Bias caused by changes in birth rate. *British Journal of Psychiatry, 115*(523), 647-657.

Harris, G.W. & Levine, S. (1962). Sexual differentiation of the brain and its experimental control. *Journal of Physiology, 181*(2), 379-400.

Heath, B.H., Hipkins, C.E., & Miller, C.D. (1961). Physiques of Hawaii-born young men and women of Japanese ancestry, compared with college men and women of the United States and England. *American Journal of Physical Anthropology, 19,* 173-184.

Helleday, J., Siwers, B., Ritzen, E.M., & Hughdahl, K. (1994). Normal lateralization for handedness and ear advantage in a verbal dichotic listening task in women with congenital adrenal hyperplsia (CAH). *Neuropsychologia, 32*(7), 875-880.

Herbert, B. (2003, November 3). Big chill at the lab. *New York Times.*

Hines, M. (2000). Gonadal hormones and sexual differentiation of human behavior: Effects on psychosexual and cognitive development. In A. Matsumoto, (Ed.), *Sexual differentiation of the brain* (pp. 257-278). Boca Raton, FL: CRC Press.

Hines, M. (2004). Psychosexual development in individuals who have female pseudohermaphroditism. *Child and Adolescent Psychiatric Clinics of North America, 13*(3), 641-656.

Hines, M., Ahmed, S.F., & Hughes, I.A. (2003). Psychological outcomes and gender-related development in complete androgen insensitivity syndrome. *Archives of Sexual Behavior, 32*(2), 93-101.

Hines, M., Brook, C., & Conway, G.S. (2004). Androgen and psychosexual development: Core gender identity, sexual orientation, and recalled childhood gender role behavior in women and men with congenital adrenal hyperplasia (CAH). *Journal of Sex Research, 41*(1), 75-81.

Hines, M., Golombok, S., Rust, J., Johnston, K.J., Golding, J., & the Avon Longitudinal Study of Parents and Children Study Team (2002). Testosterone during pregnancy and gender role behavior of preschool children: A longitudinal population study. *Child Development, 73*(6), 1678-1687.

Hoffman, H.S. & Ison, J.R. (1992). Reflex modification and the analysis of sensory processing in developmental and comparative research. In B.A. Campbell, H.

Hayne, and R. Richardson (Eds.), *Attention and information processing in infants and adults* (pp. 83-111). Mahwah, NJ: Erlbaum.

Hofman, M.A., Fliers, E., Goudsmit, E., Swaab, D.F., & Partiman, T.S. (1988). Morphometric analysis of the suprachiasmatic and paraventricular nuclei in the human brain: Sex differences and age-dependent changes. *Journal of Anatomy, 160,* 127-143.

Hofman, M.A. & Swaab, D.F. (1989). The sexually dimorphic nucleus of the preoptic area in the human brain: A comparative morphometric study. *Journal of Anatomy, 164,* 55-72.

Holt, S.B. (1968). *The genetics of dermal ridges.* Springfield, IL: Charles C Thomas.

Hotchkiss, A.K. & Vandenbergh, J.G. (2005). The anogenital distance index of mice (*Mus musculus domesticus*), and analysis. *Contemporary Topics in Laboratory Animal Science, 44*(4), 46-48.

Houtsmuller, E.J., Brand, T., De Jonge, F.H., Joosten, R.N.J.M.A., Van de Poll, N.E., et al. (1994). SDN/POA volume, sexual behavior, and partner preference of male rats affected by perinatal treatment with ATD. *Physiological Behavior, 56*(3), 535-541.

Hu, S., Pattatucci, M.L., Patterson, C., Li, L., Fulker, D.W., Cherny, S.S., Kruglyak, L., & Hamer, D.H. (1995). Linkage between sexual orientation and chromosome Xq28 in males but not females. *Nature Genetics, 11*(3), 248-256.

Imperato-McGinley, J. (2002). 5 alpha-reductase-2 deficiency and complete androgen insensistivity: Lessons from nature. *Advances in Experimental Medicine and Biology, 511,* 121-131.

Imperalto-McGinley, J., Pichardo, M., Gautier, T., Voyer, D., & Bryden, M.P. (1991). Cognitive abilities in androgen-insensitive subjects: Comparison with control males and females from the same kindred. *Clinical Endocrinology, 34*(5), 341-347.

Jamison, C.S., Jamison, P., & Meier, R. (1994). Effects of prenatal testosterone administration of palmar dermatoglyphic intercore ridge counts of Rhesus monkeys (*Macaca mulatta*). *American Journal of Physical Anthropology, 94*(3), 409-419.

Kandel, D.B. & Udry, J.R. (1999). Prenatal effects of maternal smoking on daughters' smoking: Nicotine or testosterone exposure? *American Journal of Public Health, 89*(9), 1377-1383.

Kandel, E.R., Schwartz, J.H., & Jessell, T.M. (Eds.) (1995). *Essentials of neural science and behavior.* Stamford, CT: Appleton & Lange.

Karlsen, P. & Luscher, M. (1959). Pheromones: A new term for a class of biologically active substances. *Nature, 183*(4653), 55-56.

Kelso, W.M., Nicholls, M.E.R., & Warne, G.L. (1999). Effects of prenatal androgen exposure on cerebral lateralization in patients with congenital adrenal hyperplasia (CAH). *Brain & Cognition, 40,* 153-156.

Kelso, W.M., Nicholls, M.E.R., Warne, G.L., & Zacharin, M. (2000). Cerebral lateralization and cognitive functioning in patients with congenital adrenal hyperplasia. *Neuropsychology, 14*(3), 370-378.

Kendler, K.S., Thornton, L.M., Gilman, S.E., & Kessler, R.C. (2000). Sexual orientation in a U.S. national sample of twin and nontwin sibling pairs. *The American Journal of Psychiatry, 157*(11), 1843-1846.

Kinsey, A.C., Pomeroy, W.B., & Martin, C.E. (1948). *Sexual behavior in the human male.* Philadelphia, PA: W.B. Saunders.

Kinsey, A.C., Pomeroy, W.B., Martin, C.E., & Gebhard, P.H. (1953). *Sexual behavior in the human female.* Philadelphia, PA: W.B. Saunders.

Kirk, K.M., Bailey, J.M., Dunne, M.P., & Martin, N.G. (2000). Measurement models for sexual orientation in a community twin sample. *Behavior Genetics, 30*(4), 345-356.

Klein, F., Sepekoff, B., & Wolf, T.J. (1985). Sexual orientation: A multivariable dynamic process. *Journal of Homosexuality, 11*(1-2), 35-49.

Kondo, T., Zakany, J., Innis, J.W., & Duboule, D. (1997). Of fingers, toes, and penises. *Nature, 390*(6655), 29.

Kraemer, B., Noll, T., Delsignore, A., Milos, G., Schnyder, U., & Hepp, U. (2006). Finger length ratio (2D:4D) and dimensions of sexual orientation. *Neuropsychobiology, 53*(4), 210-214.

Lalumière, M.L., Blanchard, R., & Zucker, K.J. (2000). Sexual orientation and handedness in men and women: A meta-analysis. *Psychological Bulletin, 126*(4), 575-592.

Lasco, M.S., Jordan, T.J., Edgar, M.A., Petito, C.K., & Byne, W. (2002). A lack of dimorphism of sex or sexual orientation in the human anterior commissure. *Brain Research, 936*(1-2), 95-98.

Le Vay, S. (1991). A difference in hypothalamic structure between heterosexual and homosexual men. *Science, 253*(5023), 1034-1037.

Le Vay, S. & Hamer, D.H. (1994). Evidence for a biological influence in male homosexuality. *Scientific American, 270*(5), 44-49.

Levy, J. & Heller, W. (1992). Gender differences in human neuropsychology function. In A. Gerall, H. Moltz, & I.L. Ward (Eds.), *Sexual differentiation* (vol. 11 of *Handbook of behavioral eurobiology,* pp. 245). New York: Plenum Press.

Lindeque, M., Skinner, J.D., & Miller, R.P. (1986). Adrenal and gonadal contribution to circulating androgens in spotted hyaenas (*Crocuta crocuta*) as revealed by LHRH, hCG and ACTH stimulation. *Journal of Reproduction and Fertility, 78*(1), 211-217.

Lippa, R.A. (2003a). Are 2D:4D finger length ratios related to sexual orientation? Yes for men, no for women. *Journal of Personality and Social Psychology, 85*(1), 179-188.

Lippa, R.A. (2003b). Handedness, sexual orientation, and gender-related personality traits in men and women. *Archives of Sexual Behavior, 32*(2), 103-114.

Loehlin, J.C. & McFadden, D. (2003). Otoacoustic emissions, auditory evoked potentials, and traits related to sex and sexual orientation. *Archives of Sexual Behavior, 32*(2), 115-127.

Lutchmaya, S., Baron-Cohen, S., Raggatt, P., Knickmeyer, R., & Manning, J.T. (2004). 2nd to 4th digit ratios, fetal testosterone and estradiol. *Early Human Development, 77*(1-2), 23-28.

Macke, J.P., Hu, N., Hu, S., Bailey, M., King, V.L., Brown, T., Hamer, D., & Nathans, J. (1993). Sequence variation in the androgen receptor gene is not a common determinant of male sexual orientation. *American Journal of Human Genetics, 53*(4), 844-852.

MacLean, P.D. (1990). *The triune brain in evolution: Role in paleocerebral function.* New York: Plenum Press.

Malas, M.A., Dogan, S., Evcil, E.H., & Desdicioglu, K. (2006). Fetal development of the hand, digits and digit ratio (2D:4D). *Early Human Development, 82*(7), 469-475.

Manning, J.T. (2002). *Digit ratio: A pointer to fertility, behavior and health.* New Brunswick, NJ: Rutgers University Press.

Manning, J.T. & Bundred, P.E. (2000). The ratio of 2nd to 4th digit length: A new predictor of disease predisposition? *Medical Hypotheses, 54*(5), 855-857.

Manning, J.T., Bundred, P.E., Newton, D.J., & Flanagan, B.F. (2003). The second to fourth digit ratio and variation in the androgen receptor gene. *Evolution and Human Behavior, 24*(6), 399-405.

Manning, J.T., Callow, M., & Bundred, P.E. (2003). Finger and toe ratios in humans and mice: Implications for the aetiology of diseases influenced by HOX genes. *Medical Hypotheses, 60*(3), 340-343.

Manning, J.T., Churchill, A.J., & Peters, M. (2007). The effects of sex, ethnicity, and sexual orientation on self-measured digit ratio (2D:4D). *Archives of Sexual Behavior, 36*(2), 223-233.

Manning, J.T. & Leinster, S.J. (2001). Re: The ratio of 2nd to 4th digit length and age at presentation of breast cancer: A link with prenatal oestrogen? *The Breast, 10*(4), 355-357.

Manning, J.T., Scott, D., Wilson, J.D., & Lewis-Jones, D.I. (1998). The ratio of the 2nd to 4th digit length: A predictor of sperm number and concentrations of testosterone, lutinizing hormone and oestrogen. *Human Reproduction, 13*(11), 3000-3004.

Manning, J.T. & Taylor, R.P. (2001). Second to fourth digit ratio and male ability in sport: Implications for sexual selection in humans. *Evolution and Human Behavior, 22*(1), 61-69.

Manning, J.T., Trivers, R.L., Singh, D., & Thornhill, R. (1999). The mystery of female beauty. *Nature, 399*(6733), 214-215.

Martin, J.T. & Nguyen, D.H. (2004). Anthropometric analysis of homosexuals: Implications for early hormone exposure. *Hormones and Behavior, 45*(1), 31-39.

McFadden, D.A. (1993). A masculinizing effect on the auditory systems of human females having male co-twins. *Proceedings of the National Academy of Sciences, USA, 90*(24), 11900-11904.

McFadden, D. (2002). Masculinization effects in the auditory system. *Archives of Sexual Behavior, 31*(1), 99-111.

McFadden, D. & Bracht, M.S. (2002a). The relative lengths and weights of metacarpals and metatarsals in baboons (*Papio hamadryas*). *Hormones and Behavior, 43*(2), 347-355.

McFadden, D. & Bracht, M.S. (2002b). Sex differences in length ratios from the extremities of humans, gorillas, and chimpanzees (abstract). *Hormones and Behavior, 41*, 479.

McFadden, D. & Champlin, C.A. (2000). Comparison of auditory evoked potentials in heterosexual, homosexual, and bisexual males and females. *Journal of the Association for Research in Otolaryngology, 1*(1), 89-99.

McFadden, D., Loehlin, J.C., Breedlove, S.M., Lippa, R.A., Manning, J.T., & Rahman, Q. (2005). A reanalysis of five studies on sexual orientation and the relative length of the 2nd and 4th fingers (the 2D:4D ratio). *Archives of Sexual Behavior, 34*(3), 341-356.

McFadden, D., Loehlin, J.C., & Pasanen, E.G. (1996). Additional findings on heritability and prenatal masculinization of cochlear mechanisms: Click-evoked otoacoustic emissions. *Hearing Research, 97*(1-2), 102-119.

McFadden, D. & Pasanen, E.G. (1998). Comparison of the auditory system of heterosexuals and homosexuals: Click-evoked otoacoustic emissions. *Proceeding of the National Academy of Sciences, USA, 95*(5), 2709-2713.

McFadden, D. & Pasanen, E.G. (1999). Spontaneous otoacoustic emissions in heterosexuals, homosexuals and bisexuals. *Journal of the Acoustical Society of America, 105*(4), 2403-2413.

McFadden, D., Pasanen, E.G., Raper, J., Lange, H.S., & Wallen, K. (2006). Sex differences in otoacoustic emissions measured in Rhesus monkeys (*Macaca mulatta*). *Hormones and Behavior, 50*(2), 274-284.

McFadden, D., Pasanen, E.G., Weldele, M.L., Glickman, S.E., & Place, N.J. (2006). Masculinized otoacoustic emissions in female spotted hyenas (*Crocuta crocuta*). *Hormones and Behavior, 50*(2), 285-292.

McFadden, D. & Shubel, E. (2002). Relative lengths of fingers and toes in human males and females. *Hormones and Behavior, 42*(4), 492-500.

McManus, I.C. & Bryden, M.P. (1992). The genetics of handedness, cerebral dominance, and lateralization. In Rapin, I. & Segalowitz, S.J. (Eds.), *Handbook of neuropsychology* (vol. 6, *Developmental neuropsychology*, pp. 115-144). Amsterdam: Elsevier.

Meyer-Bahlberg, H.F.L. & Ehrhardt, A.A. (1986). Prenatal diethylstilbestrol exposure: Behavioral consequences in humans. *Monograms in Neural Science, 12*, 90-95.

Meyer-Bahlburg, H.F.L., Ehrhardt, A.A., Rosen, L.R., Gruen, R.S., Veridiano, N.P., Van, F.H., and Neuwalder, H.F. (1995). Prenatal estrogens and the development of homosexual orientation. *Developmental Psychology, 31*(1), 12-21.

Micle, S. & Kobyliansky, E. (1988). Sex differences in the intra-individual diversity of finger dermatoglyphics: Pattern types and ridge counts. *Human Biology, 60*(1), 123-134.

Mittwoch, U. (2000). Genetics of sex determination: Exceptions that prove the rule. *Molecular Genetics and Metabolism, 71*(1-2), 405-410.

Moir, A. & Jessel, D. (1991). *Brain sex: The real differences between men and women.* New York: Carol Publishing.

Money, J. (2002). Amative orientation: The hormonal hypothesis examined. *Journal of Pediatric Endocrinology and Metabolism, 15*(7), 951-957.

Money, J. & Ehrhardt, A.A. (1972). *Man & woman, boy & girl.* Baltimore: Johns Hopkins University Press.

Money, J., Schwartz, M., & Lewis, V.G. (1984). Adult erotosexual status and fetal hormonal masculinization and demasculinization: 46, XX congenital virilizing adrenal hyperplasia and 46, XY androgen-insensitivity syndrome compared. *Psychoneuroendocrinology, 9*(4), 405-414.

Mortlock, D.P. & Innis, J.W. (1997). Mutations of HOXA13 in hand-foot-genital syndrome. *Nature Genetics, 15*(2), 179-180.

Mustanski, B.S., Bailey, J.M., & Kasper, S. (2002). Dermatoglyphics, handedness, sex, and sexual orientation. *Archives of Sexual Behavior, 31*(1), 107-116.

Mustanski, B.S., Chivers, M.L., & Bailey, J.M. (2002). A critical review of recent biological research on human sexual orientation. *Annual Review of Sex Research, 13,* 89-140.

Mustanski, B.S., DuPree, M.G., Nievergelt, C.M., Bocklandt, S., Schork, N.J., & Hamer, D.H. (2005). A genomewide scan of male sexual orientation. *Human Genetics, 116*(4), 272-278.

Nass, R., Baker, S., Speiser, P., Virdis, R., Bulsamo, A., Cacciari, F., Locke, A., Dumic, M., & New, M. (1987). Hormones and handedness: Left-hand bias in female congenital adrenal hyperplasia patients. *Neurology, 37*(4), 711-715.

Neave, N., Menaged, M., & Weightman, D.R. (1999). Sex differences in cognition: The role of testosterone and sexual orientation. *Brain and Cognition, 41*(3), 245-262.

Nedoma, K. & Freund, K. (1961). Somatosexual findings in homosexual men. *Ceskoslovenska Psychiatre, 57,* 100-103.

Newell-Morris, L.L., Fahrenbruch, G.P., & Sackett, G.P. (1989). Prental psychological stress, dermatoglyphic symmetry and pregnancy outcomes in the pigtail macaque (*Macaca nemestrina*). *Biology of the Neonate, 56*(2), 61-75.

O'Connor, D.B., Archer, J., Hair, W.M., & Wu, F.C.W. (2001). Activational effects of testosterone on cognitive function of men. *Neuropsychologia, 39*(13), 1385-1394.

Osborn, D.K. (1991). *Reflections on the art of living: A Joseph Campbell companion.* p. 15. New York: Harper Collins.

Paredes, R.G. & Baum, M.J. (1997). Role of the medial preoptic area/anterior hypothalamus in the control of masculine sexual behavior. *Annual Review of Sex Research, 8,* 79-87.

Paredes, R.G., Nakagawa, Y., & Nakach, N. (1998). Lesions of the medial peroptic area/anterior hypothalamus (MPOA/AH) modify partner preference in male rats. *Brain Research, 813*(1), 1-8.

Pasterski, V.L., Geffner, M.E., Brain, C., Hindmarsh, P., Brook, C., & Hines, M. (2005). Prenatal hormones and postnatal socialization by parents as determinants of male-typical toy play in girls with congenital adrenal hyperplasia. *Child Development, 76*(1), 264-278.

Pattatucci, A.M. & Hamer, D.H. (1995). Development and familiality of sexual orientation in females. *Behavior Genetics, 25*(5), 407-420.

Pawelski, J.G., Perrin, E.C., Foy, J.M., Allen, C.E., Crawford, J.E., Del Monte, M., Kaufman, M., Klein, J.D., Smith, K., Springer, S., Tanner, J.L., & Vickers, D.L. (2006). The effects of marriage, civil union, and domestic partnership laws on the health and well-being of children. *Pediatrics, 118*(1), 349-364.

Peplau, T.A., Spaulding, L.R., Conley, T.D., & Veniegas, R.C. (1999). The development of sexual orientation in women. *Annual Review of Sex Research, 10,* 70-99.

Perkins, M.W. (1981). Female homosexuality and body build. *Archives of Sexual Behavior, 10*(4), 337-345.

Peters, M., Manning, J.T., & Reimers, S. (2007). The effects of sex, sexual orientation, and digit ratio (2D:4D) on mental rotation performance. *Archives of Sexual Behavior, 36*(2), 251-260.

Phelps, V.R. (1952). Relative finger length as a sex-influenced trait in man. *American Journal of Human Genetics, 4*(2), 72-89.

Phoenix, C.H., Goy, R.W., Gerall, A.A., & Young, W.C. (1959). Organizing action of prenatally administered testosterone proprionate on the tissues mediating mating behavior in the female guinea pig. *Endocrinology, 65,* 369-382.

Pillard, R.C. & Bailey, J.M. (1998). Human sexual orientation has a heritable component. *Human Biology, 70*(2), 347-365.

Pillard, R.C. & Weinrich, J.D. (1986). Evidence of familial nature of male homosexuality. *Archives of General Psychiatry, 43*(8), 808-812.

Rahman, Q. (2005). The association between the fraternal birth order effect in male homosexuality and other markers of human sexual orientation. *Biology Letters, 1*(4), 393-395.

Rahman, Q., Andersson, D., & Govier, E. (2005). A specific orientation-related difference in navigational strategy. *Behavioral Neuroscience, 119*(1), 311-316.

Rahman, Q., Cockburn, A., & Govier, E. (2007). A comparative analysis of functional cerebral asymmetry in lesbian women, heterosexual women, and heterosexual men. *Archives of Sexual Behavior,* DOI 10.1007/s10508-006-9137-0.

Rahman, Q., Kumari, V., & Wilson, G.D. (2003). Sexual orientation-related differences in prepulse inhibition of the human startle response. *Behavioral Neuroscience, 117*(5), 1096-1102.

Rahman, Q. & Wilson, G.D. (2003a). Large sexual orientation related differences in performance on mental rotations and judgment of line orientation. *Neuropsychology, 17*(1), 25-31.

Rahman, Q. & Wilson, G.D. (2003b). Sexual orientation and the 2nd to 4th finger length ratio: Evidence for organizing effects of sex hormones or developmental instability? *Psychoneuroendocrinology, 28*(3), 288-303.

Rahman, Q., Wilson, G.D., & Abrahams, S. (2003). Sexual orientation related differences in spatial memory. *Journal of International Neuropsychology Society, 9*(3), 376-383.

Rahman, Q., Wilson, G.D., & Abrahams, S. (2004a). Biosocial factors, sexual orientation and neurocognitive functioning. *Psychoneuroendocrinology, 29*(7), 867-881.

Rahman, Q., Wilson, G.D., & Abrahams, S. (2004b). Performance differences between adult heterosexual and homosexual men on the Digit-Symbol Substitution subtest of the WAIS-R. *Journal of Clinical and Experimental Neuropsychology, 26*(1), 141-148.

Reiter, E.O. & Rosenfeld, R.G. (1998). Normal aberrant growth. In J.D. Wilson, D.W. Foster, H.M. Kronenberg, and P.R. Larsen (Eds.), *Willims textbook of endocrinology* (9th ed., pp. 1427-1509). Philadelphia: W.B. Saunders.

Rice, G., Anderson, C., Risch, N., & Ebers, G. (1999). Male homosexuality: Absence of linkage to microsatellite markers at Xq28. *Science, 284*(5414), 665-667.

Rice, G., Risch, N., & Ebers, G. (1999). Technical comment response: Genetics and male sexual orientation. *Science, 285*(5429), 803a.

Risch, N., Squires-Wheeler, E., & Keats, B.J.B. (1993). Male sexual orientation and genetic evidence. *Science, 262*(5142), 2063-2065.

Robinson, S.J. & Manning, J. (2000). The ratio of 2nd to 4th digit length and male homosexuality. *Evolution and Human Behavior, 21*(5), 333-345.

Roof, R.L. (1993). Neonatal exogenous testosterone modifies sex difference in radial arm and Morris water maze performance in prepubescent and adult rats. *Behavioral Brain Research, 53*(1-2), 1-10.

Roselli, C.E., Larkin, K., Resko, J.A., Stellflug, J.N., & Stormshak, F. (2004). The volume of a sexually dimorphic nucleus in the ovine medial peroptic area/anterior hypothalamus varies with sexual partner preference. *Endocrinology, 145*(2), 478-483.

Roselli, C.E., Larkin, K., Schrunk, J.M., & Stormshak, F. (2004). Sexual partner preference, hypothalamic morphology and aromatase in rams. *Physiology & Behavior, 83*(2), 233-245.

Roselli, C.E., Schrunk, J.M., Stadelman, H.L., Resko, J.A., & Stormshak, F. (2006). The effect of aromatase inhibition on the sexual differentiation of sheep brain. *Endocrine, 29*(3), 501-511.

Ross, J., Roeltgen, D., & Zinn, A. (2006). Cognition and the sex chromosomes: Studies in Turner syndrome. *Hormone Research, 65*(1), 47-56.

Ryan, B.C. & Vandenberg, J.G. (2002). Intrauterine position effects. *Neuroscience and Biobehavioral Reviews, 26*(6), 665-678.

Sagan, C. & Druyan, A. (1992). *Shadows of forgotten ancestors.* New York: Random House.

Sanders, A.R. et al. (1998). Poster presentation 149, annual meeting of the American Psychiatric Association, Toronto, Ontario, Canada.

Saucier, D.M., McCreary, D.R., & Saxberg, J.K.J. (2002). Does gender role socialization mediate sex differences in mental rotations? *Personality and Individual Differences, 32*(6), 1101-1111.

Savic, I., Berglund, H., Gulyas, B., & Roland, P. (2001). Smelling of odorous sex hormone-like compounds causes sex-differentiated hypothalamic activations in humans. *Neuron, 31*(4), 661-668.

Savic, I., Berglund, H., & Lindstrom, P. (2005). Brain response to putative pheromones in homosexual men. *Proceedings of the National Academy of Sciences, USA, 102*(20), 7356-7361.

Savin-Williams, R.C. & Diamond, L.M. (2000). Sexual identity trajectories in sexual minority youth: Gender comparisons. *Archives of Sexual Behavior, 29*(6), 607-628.

Savin-Williams, R.C. & Ream, G.L. (2006). Pubertal onset and sexual orientation in an adolescent national probability sample. *Archives of Sexual Behavior, 35*(3), 279-286.

Schmidt, G. & Clement, U. (1990). Does peace prevent homosexuality? *Archives of Sexual Behavior, 19*(2), 183-187.

Science (1993). *260*(5116), cover photo.

Siever, M.D. (1994). Sexual orientation and gender as factors in socioculturally acquired vulnerability to body dissatisfaction and eating disorders. *Journal of Consulting and Clinical Psychology, 62*(2), 252-260.

Silber, S.J. (1981). *The male from infancy to old age.* New York: Scribner's.

Silverman, I., Kastuk, D., Choi, J., & Phillips, K. (1999). Testosterone levels and spatial ability in men. *Psychoneuroendocrinology, 24*(8), 813-822.

Singh, D. (1993). Adaptive significance of female physical attractiveness: Role of waist-to-hip ratio. *Journal of Personality and Social Psychology, 65*(2), 293-307.

Singh, D. (1995). Female judgment of male attractiveness and desirability for relationships: Role of waist-to-hip ratio and financial status. *Journal of Personality and Social Psychology, 69*(6), 1089-1101.

Singh, D., Vidaurri, M., Zambarano, R.J., & Dabbs Jr., J.M. (1999). Lesbian erotic role identification: Behavioral, morphological, and hormonal correlates. *Journal of Personality and Social Psychology, 76*(6), 1035-1049.

Singh, J. & Verma, I.C. (1987). Influence of major histo(in)compatibility complex on reproduction. *American Journal of Reproductive Immunology and Microbiology, 15*(4), 150-152.

Skidmore, W.C., Linsenmeier, J.A., & Bailey, J.M. (2006). Gender nonconformity and psychological stress in lesbians and gay men. *Archives of Sexual Behavior, 35*(6), 685-697.

Slikker Jr., W., Hill, D.E., & Young, J.F. (1982). Comparison of the transplacental pharmacokinstics of 17beta-estradiol and diethylstilbestrol in the subhuman primate. *Journal of Pharmacology and Experimental Therapeutics, 221*(1), 173-182.

Sorenson-Jamison, C., Meier, R.J., & Campbell, B.C. (1993). Dermatoglyphic asymmetry and testosterone levels in normal males. *American Journal of Physical Anthropology,* 90, 185-198.

Swaab, D.F. & Fliers, E. (1985). A sexually dimorphic nucleus in the human brain. *Science, 228*(4703), 1112-1115.

Swaab, D.F., Fliers, E., & Partiman, T.S. (1985). The suprachiasmatic nucleus of the human brain in relation to sex, age and senile dementia. *Brain Research, 342*(1), 37-44.

Swaab, D.F. & Hofman, M.A. (1990). An enlarged suprachiasmatic nucleus in homosexual men. *Brain Research, 537*(1-2), 141-148.

Swaab, D.F. & Hofman, M.A. (1995). Sexual differentiation of the human hypothalamus in relation to gender and sexual orientation. *Trends in Neurosciences, 18*(6), 264-270.

Swerdlow, N.R., Auerbach, A., Monroe, S.M., Hartson, H., Geyer, M.A., & Braff, D.L. (1993). Men are more inhibited than women by weak pre-pulses. *Biological Psychiatry, 34*(4), 253-260.

Swerdlow, N.R., Braff, D.L., Taaid, N., & Geyer, M.A. (1994). Assessing the validity of an animal model of deficient sensorimotor gating in schizophrenic patients. *Archives of General Psychiatry, 51*(2), 139-154.

Swerdlow, N.R., Caine, B.S., Braff, D.L., & Geyer, M.A. (1992). The neural substrates of sensorimotor gatin of the startle reflex: A review of recent findings and their implications. *Journal of Psychopharacology, 6*(2), 176-190.

Tenhula, W.N. & Bailey, J.M. (1998). Female sexual orientation and pubertal onset. *Developmental Neuropsychology, 14*(2-3), 369-383.

Thompson, M.W., McInnes, R.R., & Willard, H.F. (1991). *Thompson & Thompson genetics in medicine.* Philadelphia, PA: W.B. Saunders.

Titus-Ernstoff, L., Perez, K., Hatch, E.E., Troisi, R., Palmer, J.R., Hartge, P., Hyer, M., Kaufman, R., Adam, E., Strohsnitter, W., Noller, K., Picket, K.E., & Hoover, R. (2003). Psychosexual characteristics of men and women exposed prenatally to diethylstilbestrol. *Epidemiology, 14*(2), 155-160.

van Anders, S.M. & Hampson, E. (2005. Testing the prenatal hypothesis: Measuring digit ratios, sexual orientation, and spatial abilities in adults. *Hormones and Behavior, 47*(1), 92-98.

van Anders, S.M., Vernon, P.A., & Wilber, C.J. (2006). Finger-length ratios show evidence of prenatal hormone-transfer between opposite-sex twins. *Hormones and Behavior, 49*(3), 315-319.

vom Saal, F.S. & Bronson, F.H. (1980). Sexual characteristics of adult female mice are correlated with their blood testosterone levels during prenatal development. *Science, 208*(4444), 597-599.

Voracek, M., Manning, J.T., & Ponocny, I. (2005). Digit ratio (2D:4D) in homosexual and heterosexual men from Austria. *Archives of Sexual Behavior, 34*(3), 335-340.

Voyer, D., Voyer, S., & Bryden, M.P. (1995). Magnitude of sex differences in spatial abilities: A meta-analysis and consideration of critical variables. *Psychological Bulletin, 117*(2), 250-270.

Wegesin, D.J. (1998a). A neuropsychologic profile of homosexual and heterosexual men and women. *Archives of Sexual Behavior, 27*(1), 91-108.

Wegesin, D.J. (1998b). Relation between language lateralization and spatial ability in gay men and women. *Laterality, 3*(3), 227-239.

Whitam, F.L., Diamond, M., & Martin, J. (1993). Homosexual orientation in twins: A report on 61 pairs and three triplet sets. *Archives of Sexual Behavior, 22*(3), 187-206.

Williams, C.L., Barnett, A.M., & Meck, W.H. (1990). Organizational effects of early gonadal secretions on sexual differentiation in spatial memory. *Behavioral Neuroscience, 104*(1), 84-97.

Williams, C.L. & Meck, W.H. (1991). The organizational effects of gonadal steroids on sexually dimorphic spatial ability. *Psychoneuroendocrinology, 16*(1-3), 155-176.

Williams, T.J., Pepitone, M.E., Christensen, S.E., Cooke, B.M., Huberman, A.D., Breedlove, N.J., Jordan, C.L., & Breedlove, S.M. (2000). Finger-length ratios and sexual orientation. *Nature, 404*(6777), 455-456.

Wisniewski, A.B., Migeon, C.J., Meyer-Bahlberg, H.F., Gearhart, J.P., Berkovitz, G.D., Brown, T.R., & Money J. (2000). Complete androgen insensitivity syndrome: Long-term medical, surgical and psychosexual outcome. *Journal of Clinical Endocrinology and Metabolism, 85*(8), 2664-2669.

Witelson, S.F. (1987). Neurobiological aspects of language in children. *Child Development, 58*(3), 653-688.

Woodson, J.C. & Gorski, R.A. (1999). Structural differences in the mammalian brain: Reconsidering the male/female dichotomy. In A. Matsumoto (Ed.), *Sexual differentiation of the brain* (pp. 229). Boca Raton, FL: CRC Press.

Wright, F., Giacomini, M., Riahi, M., & Mowszowicz, I. (1983). Antihormone activity of progesterone and progestins. In C. W. Bardin, E. Milgrom, and P. Mauvais-Jarvis (Eds.), *Progesterone and progestins* (pp. 121-134). New York: Raven Press.

Wyart, C., Webster, W.W., Chen, J.H., Wilson, S.R., McClary, A., Khan, R.M., & Sobel, N. (2007). Smelling a single compound of male sweat alters levels of cortisol in women. *Journal of Neuroscience, 27*(6), 1261-1265.

Yager, J., Kurtzman, F., Landsverk, J., & Wiesmeier, E. (1988). Behaviors and attitudes to eating disorders in homosexual male college students. *American Journal of Psychiatry, 145*(4), 495-497.

Yalcinkaya, T.M., Siiteri, P.K., Vigne, J.-L., Licht, P., Pavgi, S., Frank, L.G., & Glickman, S.E. (1993). A mechanism for virilization of female spotted hyenas in utero. *Science, 260*(5116), 1929-1931.

Zhou, J.N., Hofman, M.A., Gooren, L.J.G., and Swaab, D.F. (1995). A sex difference in the human brain and its relation to **transsexuality.** *Nature, 378*(6552), 68-70.

Zucker, K.J., Beaulieu, N., Bradley, S.J., Grimshaw, G.M., & Wilcox, A. (2001). Handedness in boys with gender identity disorder. *Journal of Child Psychology and Psychiatry, 42*(6), 767-776.

Zucker, K.J., Bradley, S.J., Oliver, G., Blake, J., Fleming, S., & Hood, J. (1996). Psychosexual development of women with congenital adrenal hyperplasia. *Hormones and Behavior, 30*(4), 300-318.

Index

Note: Page numbers in italics refer to terms within figures and tables.

Milton Keynes UK
Ingram Content Group UK Ltd.
UKHW031135141024
449569UK00006B/176